景观启示录
——吴欣与当代设计师访谈

The New Art of Landscape
— Conversations between Xin WU and
Contemporary Designers

吴 欣 编著

中国建筑工业出版社

前言
——复原景观设计的艺术性

在全球化迅猛发展的今天,景观设计向何处去?它是否还在延续着伊恩·麦克哈格和美国景观设计师协会所指定的生态路线?还是已经在寻找新的方向,新的目标,新的美学?

自 Landscape Architecture 学科在世纪之交在哈佛设立以来,已经有越来越多的国家和大学建立了自己的景观设计和研究院系。这不仅是国际交流和现代教育的必然,也是各方对一些全球性的挑战(如环境和可持续性发展问题)的共同回应。新时代的设计师们——许多来自于景观建筑学以外的背景,另外一些(如中国和亚洲设计师)来自于有悠久文明的景观传统——已经在探索新的领域,重新定位景观设计。本书的目的就在于发掘和介绍这样的一些景观设计的最新动态。

这个想法是在应邀为北京大学《景观设计学》杂志撰写"当代景观评论"专栏(中英双语,双月刊)的过程中逐渐形成的。在 2009 年和 2010 年,专栏刊出了十二期两个系列:系列一是关于建成项目的评论;系列二是对六位来自于中国、欧洲和美国的景观设计师和艺术家的访谈。这些访谈(增补一篇[①])在此处结集出版。初看起来,我所选择的这七位被采访者——俞孔坚[中国]、戴安娜·巴摩里[美国]、贝尔纳·拉素斯[法国]、帕特里夏·约翰逊[美国]、埃里克·董特[比利时]、林璎[美国]、帕奥罗·伯吉[瑞士]——是截然不同的。他们的作品和他们的观点及方法论一样,天差地别。然而,重温他们的话语,我们会发现他们之间一种深层的共性。那就是,他们不仅都努力地与商业化设计保持距离,而且都决心复原景观设计作为一种对当今社会有建设性意义的艺术。

在西方,景观设计诞生于一个偶然的事件:18 世纪 30 年代英国伯林顿伯爵三世理查·波耶勒(1694–1753 年)请受过学院派历史绘画训练的威廉·坎特

[①] 增加的林璎访谈原来发表于另一本中国设计杂志《风景园林》(2010.01) 16–43 页。

(1685—1748年)为他设计克斯韦克花园。就这样，坎特"越过了樊篱"，开始了一个统合花园和景观的艺术革命①。自那时起，设计艺术逐渐地从对阿卡迪亚式（希腊神话中田园牧歌的极乐世界，相当于中国的世外桃源）景观的追求，转向了多种对自然画意论模式的表现。这种趋势从18世纪下半叶一直延续到19世纪和20世纪前期。第二次世界大战以后，伊恩·麦克哈格理论的提出，进一步地将景观设计推离了视觉艺术而转向生态规划。也就是在同一时期的20世纪70年代初（距坎特的时代两个半世纪以后），少数先锋艺术家对景观艺术重新产生了兴趣——如英国新阿卡迪亚人组合中的伊安·汉密尔顿·芬利（1925—2006年）和安迪·高斯沃斯（1956—），国际境况主义者组合中的荷兰艺术家路易·G·勒华（1924—），美国大地艺术的倡导人罗伯特·史密森（1938—1973年）。然而，他们中没有一个人试图完全地再定义景观设计的整体架构。

要了解景观设计的新方向，我们必须把目光转向本书中介绍的景观设计师和艺术家们。他们不仅致力于复原景观设计的艺术性，而且赋予了它新的意义和美学②。他们中的好几位是由视觉艺术自发地转向景观设计的。贝尔纳·拉素斯对科西嘉岛民居色彩的研究，帕特里夏·约翰逊在上纽约州所创造的大型室外色彩雕塑，林璎的华盛顿越战纪念碑，都是强有力的例证。从不同的艺术角度出发，他们都把景观作为一个艺术创造的宝库。同时，从事相关的设计领域的俞孔坚、戴安娜·巴摩里、埃里克·董特、帕奥罗·伯吉，又都发现了艺术对设计创意的价值。当然，每一位访谈对象是从完全不同的艺术中找到启迪的：绘画、雕塑、动感艺术、光画、抽象艺术、大地艺术、平面设计、装置、舞蹈、音乐……不一而足。如果本书能收入更多的设计师和艺术家，这个单子还会更长。这对意欲从

---

① 贺瑞斯·沃尔普(1717—1797年)对威廉·坎特的赞美值得在此引用："他越过了樊篱，把整个自然看作一个花园。他感觉到山川谷地间微妙的对比和完美的过渡，品赏地形起伏的美丽，指出独立的树丛能何等轻易地烘托和点缀主体。"

② 这远远不是一个完全的名单，其他许多人都为此发展作出了贡献。然而本书中所录入的设计师和艺术家的多面性揭示了当今正在发生的探索的深度和广度。由于本书篇幅的限制，我很遗憾不能收入以下几位我敬佩并曾评论过的设计师：斯蒂格·安德森[丹麦]，费尔南多·查塞尔[巴西]，安迪·曹[美籍越南裔]，中村良夫[日本]，穆罕默德·夏西尔[印度]。

艺术中发掘灵感的人们无疑是一个巨大的鼓励。

在20世纪80年代，彼得·沃克已经在某种程度上早于这一群体将艺术引入了设计。我们既不应该忽视沃克的贡献，又不应该将他与目前这个领域里的最新动向混淆起来。沃克以一种形式主义的姿态，来抗衡当时景观建筑学中由艺术转向纯环境规划的强势。他提倡设计要为场所带来一种形式化的识别性，强调设计的平面性和图案的重复性。这是一个极其有力的思路：与此前所有的景观设计理念不同，它不是基于一种现成的形式语言，而是鼓励设计师发明一个新的景观语系。这与赖特坚持他的每一个Usonian住宅都要有独特的建筑和雕塑的想法相似。同时，沃克对平面性和重复性的强调，为他进一步在景观中引入大地艺术风格的简洁单体提供了背景；这在某种程度上与法国17世纪花园设计大师勒·诺特在Chantilly设计的以雕塑点缀的花坛和水池同出一辙[①]。沃克主张中更重要的一点要归功于艺术评论家克莱默·格林伯格（1909-1994年）所倡导的现代绘画以平面性反三维错觉传统的理论；而格林伯格理论的提出是与西方绘画史上亨里奇·沃夫林（1864-1945年）关于布面油画特性的论断相对应的。因而，这种形式主义的姿态的要旨是在于发现超越人类创造性的抽象艺术的"本质"。

本书中所介绍的景观设计师和艺术家们有着完全不同的见解和目标：

超越形式主义：为新的景观体验而设计

与沃克不同，本书中所有的被采访者都坚持要创造一种存在于使用者互动中的艺术。这种艺术的质量来自于它所能产生的观众反应和经验的多样性。他们的艺术深切地关注人与自然割裂的问题；但又不是一种被动的补救，而是重在激发新的景观体验和新的文化取向——这一点在俞孔坚所用的强烈而具有挑战性的手法和林璎的纪念碑中，能得到绝佳的证明。与人的体验交流有无数种方式，这些设计师各自从完全不同的角度进行了探索。比如林璎就指出"照片永远都无

---

① 彼得·沃克自己已经解释过他是如何在带学生参观Chantilly花园时顿悟的，当时他正在苦苦寻找一条将20世纪六七十年代的美国战后艺术用到景观设计中的途径。

法传递创作的意图。你必须通过体验来理解这些作品"。她把对她的作品的体验比作阅读一部私人回忆录——"一种复杂而感性的行为"。这就意味着这种新艺术志在引发一种艺术家本身并不参与的、作品与观众间的互动。帕特里夏·约翰逊和戴安娜·巴摩里视她们在特定场地上的设计为诱饵,引发如环境美学家阿诺德·柏林特所称的"与自然的能动交融"。她们把艺术定义为创造自然新体验、将想象力导向思考人作为生物链中普通一环的工具。贝尔纳·拉素斯、埃里克·董特和帕奥罗·伯吉也都指出他们的作品注重于形式之外的个性化情感、认知和理解。这是一种目的在于使观众体验到在自然中的顿悟的艺术。伯吉说:"我要提醒人们,在我们可见的事物之外还有许多东西;使人们稍稍更好奇一点,更愿意询问:在地平线后面藏着什么?"因而,使他们的设计成为艺术的不是形式和构图中的某种"本质",而是不同人群所获得的景观体验的多重性。这一在哲学层面上从本质到体验的巨变,使这种新的景观艺术与中国哲学更为靠近。

这一新的景观艺术是一种导向性的艺术。它们营造一个想象的、理想化的景象,或者是提供一个观赏性的形式和构图,这些新景观的创造者们把他们设计的重点放在将人的注意力导向设计以外的自然万物。俞孔坚的泡泡花园和帕特里夏·约翰逊的几乎所有作品就是这样。在设计中,他们拒绝将形式和意义强加于观者,而是力求创造尽可能复杂的感性经验。与伊安·汉密尔顿·芬利的探索相似,林璎把阅读文字看作是人的第六感,认为它与其他感官结合可以引发强烈的情感对话。在她的作品中有许多神秘和含糊的东西,这正是因为她在探索感知的边缘区;我们被迫去发现作品以外的情景,启动个人深层次的体验。其他设计师们用各自的方式来引导观者经验,将他们诱入一个舒适的处境,然后思考一些偶然性的自然体验。彼时,观者顿然,悟到艺术的真谛。因而,这种新的艺术只存在于观众个体的接受中,但同时是有目的而不迷惑的。景观设计导向人类的想象力以重新唤醒我们对自然的关注、梦想,并共享梦想。这是一种人与万物之间新型的交流。

扩展的创造领域:城市和所有公共基础设施

与当代艺术中面向博物馆和评论家的精英主义态度不同,新的景观艺术是为大众的，为日常生活中的生活体验的。它意在改变我们的整个生活世界，人迹所即之处使自然成为新的奇迹。这就为设计者提出了一系列有趣的挑战。越来越多的人住在城市里，而大多数的城里人必须靠公共基础设施才能接近自然。正如戴安娜·巴摩里所指明的："公共基础设施是具有些许自然特性的工程系统，其重要的功能在满足人类生活的需要。"这就要求我们将自然带入到所有人类活动触及的地方——停车场、马路、港口、垃圾堆场、水处理厂、废弃矿山，等等。一如帕特里夏·约翰逊从她1969年的"家园与花园"方案以来所一直追求的。反过来，在基础设施设计中融入新的自然观将激起人们对自然的新兴趣。艺术有可能引发关于城市文化的大讨论和新发展。比如说，贝尔纳·拉素斯就以他的高速路、休息区、高铁和机场景观设计为大众所熟悉；而俞孔坚则以他的反规划理念知名。

然而，这些新方向也使景观创造者面临一个悖论。一方面，他们的设计是将人与自然阻隔开来的现代城市和技术的一部分——传统上，城市和基础设施是体现人与自然分离的人工构筑物，支持认为人和万物有根本区别的西方理念。现代技术和应用科学正是建基于这种人与万物的对立之上的。城市生活依赖于技术，现代居住的目的就在于保护人类，将人从自然界中孤立出来，与人类以外的万物保持距离（除了少数选作食物和玩物的动植物）。这就意味着，在本书设计师和艺术家们的眼里，我们现有的整个城市环境必须被彻底地改革。另一方面，这样的巨变是不可能在一夜之间完成的，只能循序渐进。深层次的变革只会在大量市民认识到景观的道德纬度并下决心改变时发生。从这里，我们的被采访者们看到了景观艺术的新作用。不同于现代艺术先锋和环保主义者，这些设计师和艺术家们从不声称他们能设计一个人与万物和谐共处的完美未来。而是力图激发广大群众面临挑战的急迫感（比如林璎的新作《什么正在消失？》）。

新美学：能动交融和异质多义性

尽管不同人群的文化差异，我们中的多数人对自然万物的存在视而不见。同时，文化的差距阻止人们走到一起。新的景观艺术旨在创造引人注目的设计，这种设计为大量的、有不同期望值的人们提供意料之外的体验，而且使他们可以一起共享与自然的能动交融。

在许多不同的情况下，人们都可能被带入如此这般的新体验中。埃里克·董特解释了他是如何将历史性花园解读为文化和自然双遗产，并利用花园的文化纬度来帮助人们发现其自然纬度。谈到荷兰艺术家路易·G·勒华的生态大教堂，董特特别注意到勒华"面对自然，何时有为，何时无为"；伟大的景观往往归功于人类的无为而治，而且没有定型。一位西方当代艺术家与老子不谋而合，多么神奇！董特的客户们都有很高的文化修养，对艺术有各自的喜好和眼光，并不一定倾向于与自然的能动交融。所以他利用他的设计作为引子，将观者引向花园中不受人为影响的自然部分。他沉思道："对我来说建立一个与自然和谐的良好平衡很重要。比如说，可以设计50%，而将其余的50%留给自然。"这句话当然不是设计的成法，但它提出了一个有趣的论题：什么样的艺术能创造一个自然得以充分展开的场所？本书中的每一位设计师和艺术家自有不同的方法，但他们都同意自然不是一个永恒不变的本质。恰恰相反，他们把自然看作一个变化无穷的存在—万物。不再像18世纪设计师那样追求田园牧歌的完美，将景观看成是对过去乌托邦式的和谐的体现，他们把景观视为一个动态的、异质的世界。

他们关注人类在自身进化过程中与万物的能动交融。他们的作品中没有传统建筑构图中那种理想化自然同位统一的恢弘。与其注重于一种平衡完美的表象，这些当代景观创造者强调差异性、生物多样性、文化的不连续性、不同位性和生物链（包括人类）。所有这些都使他们设计的景观成为恒变的活的整体。他们为自然生命的异质多义性而设计，为人与万物的能动交融而设计。自然是不可表现的，但自然的存在可以通过景观设计表现得更加显著。

再现历史而不怀旧

现代主义者和环保主义者都不太看重历史。现代主义者感兴趣的是创造一个与工业化之前的过去无关的新世界，环保主义者感兴趣的则是一个免于污染的新世界。然而，目前对于一个日新月异的世界的焦虑诱发了大量的怀旧情绪，使人们反省历史对了解现在的重要性。自然界，尤其是人文自然，有其自己的历史。

本书中的设计师和艺术家们从不同的角度研究了景观艺术和历史的关系。当他们将目光投向过去，他们看到人与自然万物的互动在景观中刻下了不同空间和生态组成的痕迹，写下了对人的生存环境的不同文化理解。他们中的几位对场地上的人文遗迹感兴趣，将其视为人与景观之间在不同文化下异质互动的佐证。另几位转向场地上植物和动物的生命史，还有几位将原始植被看成是对历史价值的反馈，或者在文化史中寻求活在当地人记忆中的传统。他们中的许多表达了对今天环境意识日益高涨的人们所面临的各种难题的关注。与民俗主义者和复古主义者不同，这些景观创造者中没有一位试图再造一个幻象式的过去或者怀旧的场所。他们宁可在对过去的回忆上创造出一种向前看的境界，使人们正视时光的流逝，面向未来。这就提出了关于在设计中如何通过历史来表现景观自身发展的问题。本书中的设计师们选择构建诗意的抽象自然的方法。用贝尔纳·拉素斯的话来解释就是："一个艺术家不可能提供对自然直接的模仿。一旦他意识到他不可能再现自然，他就必须反省他的作品与自然之间的鸿沟。我认为艺术在很大程度上是对存在于直接模仿之外的边缘领域的发现。"

  历史的挑战在有着璀璨历史，但滞后于现代化进程的文明古国（如中国、波斯和印度）尤其巨大。这些设计师和艺术家们的真知灼见必然会为我们衔接历史与当今提供新的思路和视角。我已经提到了拉素斯将历史上的文化多样性作为一个场所上存在的景观多样性的源泉，在设计中力求揭示场地内在的不同历史性景观层面。林璎走了一条不同的路。这里我要简单地提一下著名的、但她在访谈中不愿多谈的越战纪念碑。

  初看起来，那两片插入土地的巨大的黑墙并不怎么表达多样性。但是我要分析一下，这个项目如何体现了典型的西方概念中的多样性并同时激发一种倾向于亚洲文化的态度。这个纪念碑被设计成一个"反－纪念碑"，与它的两面V－型墙所指向的另外两个纪念碑（华盛顿纪念碑的方尖塔和林肯纪念堂）截然相反。它通过典型的后抽象表现主义艺术手法来与周围众多的新古典主义的纪念性建筑区分开来，为华盛顿国家广场带来了一种意想不到的异质性。与其高耸出地面，用巨型的建筑体量来标示一个与大众分离的个体，越战纪念碑沉入地下来纪念在战争中阵亡的军人们。但它根本不像坟墓;而是有着一种田园式的非建筑的特点。在原初的设计中,参观者是通过从广场斜向墙根的草地缓坡来接近纪念碑的。这片大草地本身毫无疑问地表征了众所周知的"美国大草坪"——美国梦中自然之家的象征。尽管"美国大草坪"是一个文化的象征性的自然,大多数美国人却把它看作他们热爱自然的标志。这是一种高度浓缩的、人与自然超凡相遇的象征,一如美国先验主义的鼻祖拉尔夫·沃尔多·爱默生(1803－1882年)所歌颂的那样。因此,尽管似乎有点自相矛盾,草地的单一性在20世纪美国的文化背景下其实成为了自然的多样性代表。越战纪念碑的设计是一个巨大的成功,参观者众多,走在草坪上,以至于毁坏了它的完美。这个问题不得不通过增加步道来解决;石砌步道紧贴着刻有58195阵亡将士姓名的纪念墙,墙上名字的排列不分性别、人种、民族。我们可以说这似乎是又一次对异质性的忽视。然而,在欧洲的两次世界大战美国阵亡将士公墓都是由锥形无名墓石组成的,通过阵亡士兵的数量(而非个体)来象征美国牺牲精神的神圣。越战纪念碑却反其道而行之,每个阵亡将士都以姓名列出。当参观者沿墙踱步会看到他们自己的影像反映在千百阵亡者名录中;毫无疑问地,这里会产生出一种紧密的双关性——亿万美国人对越战纪念碑的祭奠,与西方典故中荷马、维吉尔、但丁等描写过的逝去的英杰的云集的地下伊利生界。每个美国参观者在此重复了尤利西斯、安尼斯或者但丁自己的所为,试图从居住在斯蒂克斯河对岸的伊利生界的亡灵中寻找先人。姓名指向死者漂无的灵魂。在每个士兵看似一致的表象后面,亲人们会在反映在墙上的众多影像和姓名之间找到与他们灵魂相通的那一个。这样的经验使每一次的参观都成为独一无二的经验。管理方鼓励人们从墙上拓下亲朋的名字,并拍照留念。这种无法稍减的亲属与普通参观者之间的区别使从墙上拓写姓名成为一种仪式。尽管林璎的设计并没有有意识地借用中国文化,这样一种生者和死者之间的经由触觉的交流,使人联想起中国传统中的清明节扫墓和碑石拓片。

越战纪念碑是一个使人们对异质多义性的诉求强化的艺术作品。有些退伍军人觉得这个反-纪念碑没有充分彰显他们为国家所作出的牺牲,成功地要求在场地上增加了两组雕塑。他们没有理解到纪念碑中所蕴含的丰富性,或者是他们仍希望这个纪念碑追随第二次世界大战后的雕塑传统。根据他们的要求增加的部分，其实落入了美国文化的巢臼——性别、人种和民族的差异在这些超人尺度的群雕中被高度象征化。新的雕塑充其量是提供了一个墙上已经富含的异质多义性的苍白无力的影像，可以将它作为对此种异质多义性的一个无意识的回应。简而言之，越战纪念碑的整体为华盛顿国家广场的纪念性景观引入了异质多样性，它衔接美国近代史、欧洲文化、亚洲传统和当代艺术理念。这是一个从历史和文化传统中吸取营养的全新的创造。从这个角度来说，林璎在访谈的最后号召中国设计师在新旧之间寻找平衡，其实是鼓励发掘一种对历史和文化高度敏感的创意。

被访谈的所有设计师和艺术家都表达了对有中国创意的新景观的期望。我曾请每一位被采访者给中国设计师一些建议。出于对中国文化的尊重，他们中没有一个提到了具体的方法；但都呼吁中国同行们探索一种新的景观设计，既反映中国几千年的山水文化，又正视当代特有的问题。这也许看似一个不可能的任务；然而，贝尔纳·拉素斯就在他的先锋派的COLAS基金会花园和苏州古典园林间发现了共性："苏州花园看起来和我的作品完全不一样。但它们对氛围的营造，间接而又意味深长，绝对是与我的作品相通的。"如此这般的卓见邀请中国设计师们故地重游，以一种当代的视角来重新发现中国园林艺术，包括多重感性、氛围、非表象性、无为而治、交流互动、片断性、不连续性、不统一性、人与万物的关系，和中国艺术（包括书法和赏石）的构图法则，等等。

最后，我要特别感谢《景观设计学》杂志社和中国建筑工业出版社（特别是佘依爽和郑淮兵两位编辑）的支持，使这些访谈能与更多的读者见面。从个人的角度来说，我格外地高兴这本书出版在中国这样一个有着世界上最长的景观文化的文明古国。景观文化在今天的中国完全可能比以前的任何时代都丰富和生动。

吴欣

2011年春末

于寂然不动轩

Preface:
Restoring Landscape Design as An Art

Where is landscape design heading in this era of globalization? Does it still follow the domain of ecological engineering along the lines prescribed by Ian McHarg (1920-2001) and the ASLA? Or, does it grope for new directions, new senses of purpose and new aesthetics?

Since the establishment of Landscape Architecture as a discipline in Harvard at the turn of the 20th-century, more and more countries and universities have developed departments of landscape design and studies. This is not only a result of international exchange and modern education, but also responses to shared global issues, such as environment and sustainability. New age designers—many with training other than landscape architecture, others (like the Chinese and Asian designers) with a rich heritage of engagement with landscape—have continued to test new grounds and to redefine the field. To explore and present a few of such new directions is the goal of this book.

The idea surfaced during the course of writing a bi-monthly bilingual column of "Contemporary Landscape Criticism" at the request of the journal of *LAChina* (Peking University). In 2009 and 2010, 12 issues of the column have been published in two series—first, a review series on built projects; and second, an interview series with 6 selected landscape designers and artists from China, Europe and the USA, which is presented here in this book with one addition. [①] At first sight, these 7 designers—Yu Kongjian, Diana Balmori, Bernard Lassus, Patricia Johanson, Erik Dhont, Maya Lin and Paolo Bürgi—are extremely different from one another, and the works themselves are as varied as their perspectives and approaches. And yet, as one re-reads their words, one senses a deeper commonality. This is not only a shared effort to disentangle themselves from the taken-for-granted furrows of business-oriented design, but also a

---

[①] The one added interview by me with Maya Lin was originally published in a different Chinese design journal: *Feng jing yuan lin* (2010.01): 16-43.

will to restore landscape design as a meaningful art in contemporary societies.

In the West, landscape design was born out of serendipity when William Kent (1685-1748), an artist trained in academic history painting, was called upon by Richard Boyle (1694-1753), 3rd Earl of Burlington, to design the gardens of Chiswick in the 1730s. This is the time when Kent "leapt the fence", and engaged an art revolution that led from garden into landscape. [1] Since then landscape design moved away from an artistic evocation of Arcadia, to various representations of the picturesque nature during the late-18th, the 19th- and the early-20th century. With Ian McHarg, in the post-WWII era, it moved further away from the visual arts to ecology. It was also in this same period, at the turn of the 1970s two and a half centuries after William Kent, that a few artists again showed great interest in exploring landscape as an art medium—one might mention the New Arcadians in Great Britain with Ian Hamilton Finlay (1925-2006) and Andy Goldsworthy (1956-), the Situationist International with Louis G. Le Roy (1924-) in the Netherlands, and initiator of the Earthwork movement Robert Smithson (1938-1973) in the USA. However, none of them made a serious attempt at redefining the field of landscape design as a whole.

For new directions, one has to turn to the selection of landscape designers and artists presented in this book to find figures whose work aim at restoring landscape design as an art, at the same time to give it a new sense of purpose and new aesthetics. [2] Several of them went through a serendipitous experience that led from the visual arts to landscape design. Bernard Lassus' survey of the use of color in vernacular

---

[1] Horace Walpole's (1717-1797) words in praise of William Kent (1685-1748) are well worth quoting again: "He leapt the fence and saw that all nature was a garden. He felt the delicious contrast of hill and valley changing imperceptibly into each other, tasted the beauty of the gentle swell, or concave swoop, and remarked how loose groves crowded an easy eminence with happy ornament."

[2] This is far from an exhaustive list, since many other designers have also contributed to this development. Yet the diversity of these contributors demonstrates the breadth and depth of the renewal taking place at present. Due to the size of this book, I regret that I cannot include a few more designers whose works I admire and have reviewed: Stig Andersson (Demark), Fernando Chacel (Brazil), Andy Cao (Vietnam-US), Yoshio Nakamura (Japan), and Mohammad Shaheer (India).

architecture in Corsica, Patricia Johanson's creation of a painted outdoor sculpture in upper state New York, Maya Lin's design for the Vietnam Veterans Memorial in Washington DC, all offer a case in point. Starting from different artistic perspectives, they all saw the potential offered by landscape as a domain of creative art. At the same time, Yu Kongjian, Diana Balmori, Eric Dhont and Paolo Bürgi all with training in design were discovering the value of the arts for significant breakthrough in their own works. Of course, each of these interviewees had different arts as references-painting, sculpture, kinetic art, light painting, abstract art, earthworks, graphic design, installations, dance, music, just to name a few. The list would be extended had we introduced more contemporary artists and landscape designers. It provides a rich encouragement for contemporary landscape designers who would like to turn for inspiration to the arts.

In the 1980s, Peter Walker had already responded to the arts somewhat earlier than this group of designers. We should neither overlook Walker's contribution, nor confuse it with the present developments in this field. To the radical turn of landscape architecture away from the visual art towards environmental planning, Walker opposed a formalist attitude. He advocated designs that impose a formal identity upon a place, asserting flatness and repetition of the pattern. His was a very powerful idea, which, contrary to all previous garden art, did not rest on the development of a stylistic vocabulary but invited designers to invent a new vocabulary of the landscape. It reminds of Frank Lloyd Wright (1867-1959) insistence on creating a specific architecture and sculpture vocabulary for each of his Usonian houses. Walker's assertion of flatness and repetition of patterns on a landscaping unit provided him with a background on which to introduce further works of art reminiscent of earthworks, in a way that parallels the parterres and water basins decorated with statues by the French 17th-century garden designer Le Nôtre (1613-1700) at Chantilly. [1] A most important aspect of his doctrine however

---

[1] Peter Walker himself has often explained how he has been inspired by a visit to Chantilly with a group of students, at a time when he was groping to find a translation of American contemporary art of the 1960s and 1970s into landscape architecture.

is deeply indebted to Clement Greenberg's (1909-1994) call for flatness in modernist painting against three-dimensional illusion, in defense of the specificity of canvas painting along the lines established by Heinrich Wölfflin (1864-1945) in the history of painting. Such formalist attitude aims at discovering the essence of art abstracted from human artistic inventions.

The claims of the landscape designers and artists in this book are of a very different sort:

Beyond Formalism: Design for New Experiences of Nature

To the contrary of Walker, the landscape designers presented here all insist on producing an art that comes into being through interaction with its users, an art that derives its qualities from the diversity of responses and experiences it affords to its audience. Their art is deeply concerned about the alienation of humans from nature. Yet it is not a remedial art, but an art that aims at new experiences of nature and at stimulating new cultural attitudes—which is certainly demonstrated by Yu Kongjian in a most strong and provocative way, and by Maya Lin in her memorials. There are very different ways of addressing human experience, and these artists each are exploring them from various ends. Maya Lin for instance remarks about her works that "Photographs never convey the idea of the work. You have to experience these works to understand them." And she likens the experience of her work to a reading of personal memories, "a complicated and physically emotional act." This implies that this new art is about triggering a response which she, as the artist, cannot anticipate. Patricia Johanson and Diana Balmori see the work they design on a site as an initiator towards an "engagement with nature", to borrow the term coined by aesthetician Arnold Berleant. Thus they claim that their art is about creating a new experience of nature by framing people's capacity to imagine themselves as part of the great chain of living beings. Bernard Lassus, Paolo Bürgi and Eric Dhont also expect their works to suggest personal emotions, perceptions and understandings which go beyond any intention attached to particular objects in their design. This is an art of landscape that aims for the audience to experience an epiphany of nature. Bürgi puts it: "I wanted to remind people that there is something beyond what we see, to make them a little bit more curious and willing to ask 'what is hidden beyond the horizon'." Thus it is not the unique essence contained in the form or composition that makes the landscape a work of art, but the variety of experiences that it procures to different people. This is a major philosophical shift from essence to experience, that brings this new art of landscape closer to Chinese philosophy.

The new art of landscape is an art of framing. Instead of offering representations of some imagined or ideal landscape, or trying to propose a form that would be appreciated for its own composition, these contemporary creators of a new landscape art aim at directing attention to nature, beyond the objects that they design. Such is the case in Yu Kongjian's Bubble Garden and in all works by Patricia Johanson. They avoid imposing an object or a meaning to their audience, and they seek to create a level of sensorial complexity as large as possible. Maya Lin, following a path already trodden by Ian Hamilton Finlay, insists that reading is the 6th sense, and that it gains in emotional intensity when conjoined with the other senses. So there is something elusive and mysterious about her work since it explores the edge of perception, it demands to look beyond the object and the situation, to engage in some deeper and personal exploration. Other landscape creators adopt other ways of framing the mind of their visitors, of luring them into a pleasurable experience, and leading them to ponder about a sudden experience they had not anticipated to which this pleasure leads them. Only then, in a moment of pondering what lies beyond their immediate perception, do they reach a full experience of this art. Thus this art only exists thru personal reception, but it is also always about something more focused than puzzlement. It frames human imagination in order to re-awaken humans to the presence of nature, to invite them to dream and exchange their dreams about a new compact between humans and non-humans.

An Expended Domain of Creation: the City and All Infrastructures

Contrary to the elitism of much contemporary art created for museums and critics, this new art addresses the public at large and aims at transforming the everyday life experience of nature by groups of people. Hence, it aims at transforming the living world, at making nature an object of new wonder wherever human being congregate and dwell. This sets an interesting array of challenges for landscape designers. Since more and more people live in cities and the majority of city-dwellers reach into nature through infrastructure, as Diana Balmori articulates, "infrastructures are engineering systems availing themselves of some aspects of nature," which "fulfill an important function for sustaining (human) life." It calls for bringing a new experience of nature to people wherever human activities reach—parking, roads, harbors, garbage dumps, water treatment plants, deserted mining sites, etc; just as Patricia Johanson has pursued since her 1969 House & Garden drawings. Besides, embodying new attitudes towards nature in our infrastructures should stimulate a renewal of attention to nature. It may trigger debates and the development of new trends in city culture. Bernard Lassus for example is famous for his landscape designs for motorways and their rest areas, rapid train infrastructure and airfield, while Yu Kongjian is known for his idea of negative-planning...

Yet these developments also confront landscape designers with a paradox. They place their interventions in city or technical contexts that powerfully predicate the alienation of humans from nature: cities and infrastructures are historical artifacts that embody and naturalize the modern belief in the separation of human and nature, and sustain the western idea of an essential difference between humans and non-humans. Technology and empirical science are predicated upon this opposition between reflective humans and non-reflective others. Modern urban life is pervaded by technology, and modern habitation aims at protecting human from nature, at isolating humans from the environment, at keeping non-human living beings at bay (merely selecting a limited number of plants and animals for food and leisure). In the developed world all objects around city-dwellers as well as their activities and behaviors are accustomed to be predicated upon the distinction between human and nature. This implies that, in the eyes of the landscape designer in this book, our current urban environment has to be radically transformed. Yet, such sea-change cannot happen overnight, and it has to be incremental. Deep reforms will only take place when large numbers of people consider it as an ethical must and develop common perspectives to achieve it. This is where our designers see a new role for landscape art. Contrary to the avant-garde artists of modern art and to the landscape architects of the environmental movement, these designers and artists do not claim they can prefigure the world in which humans and non-humans could live in a sustainable way. Instead, they aim at stimulating the urgency of facing the challenge, like in Maya Lin's new project *"What is Missing?"*

A New Aesthetics: Engagement and Heterogeneity

Despite of cultural differences between groups of people, most of us turn a blind eye to most aspects of nature and non-human life around. Yet, cultural differences have to be acknowledged because they prevent people from getting together. The new landscape art is bound to create attractive objects that bring a radically unexpected experience to large number of people with very different expectations and yet enable them to share a way of engaging with nature.

There are many different circumstances under which people may be brought to such a new experience. Erik Dhont for instance explains how he looks at a historical garden as both a cultural and a natural heritage, and designs for people who are already attracted to the cultural dimension of a garden in order to help them discover its natural heritage. Commenting on the eco-Cathedral by Dutch artist Louis G. Le Roy, Dhont notes that Le Roy "explained how he would do things, and also avoid doing things." The great landscape derives from non-action and has no form. What a surprise to observe how contemporary landscape artists in the Western world are crossing the paths of the philosophy of Laozi! Dhont knows that his clients are highly cultivated, having specific demands and expecting art forms to confirm their own cultural inclinations, which are not necessary leading towards a direct engagement with nature. So, he uses his design as stepping stones that guide garden-goers towards places where he has allowed untamed nature to thrive. He ponders that "it seems to me…important to establish a good balance in coordination with nature. For instance, you may design fifty percent of the project and leave the other fifty percent to nature." This sentence certainly is not a law for design, but it raises an interesting issue: what is the art of creating a place where nature thrives? Here again, each of the designers in this book adopts a different approach. Yet they all agree implicitly that there is no essence of nature. To the contrary they see nature as infinitely varied. Instead of seeing landscape as an embodiment of a past or a utopian harmony, as the Arcadian landscape designers of the early 18th c did, they see landscape as a heterogeneous and changing world.

They are concerned with the willful engagement of humans in the course of life evolution. So their works shun the grand tradition of architectural composition of isotropic space with successions of pictures of an ideal nature. Instead of aiming at the perfection of a balanced representation, these contemporary landscape creators stress differences, varieties of biotopes, cultural discontinuities, anisotropies, intensity of predatory relationships (including by human beings). All these make their landscapes into living entities, for ever on the move. They design for heterogeneous presence of natural life, for the diversity of ways of exploring, interacting and engaging. Nature cannot be represented, but its presence can be intensified.

Vindicating History against Nostalgia

Neither modernism, nor environmentalism has shown much attention to history. Modernists were keen on creating a brave new world that would be freed of any tradition of its pre-industrial past, and environmentalists a world freed of industrial pollution. Yet the anxiety with respect to the consequences of the present course of change in the world has unleashed many forms of nostalgia and a growing awareness of the importance of history for understanding the present. Nature, and even more cultural landscapes, have their own history. This group of landscape designers has been engaged in very diverse ways in the expression of a relationship between landscape art and history. When they turn an eye to the past, they can see that interactions between humans and non-humans in the landscape have yielded different forms of spaces and ecologies, as well as different cultural understanding of the human condition. Some of them are interested in traces of human activities on the sites as testimonies of different cultural understanding of the relationships between human and landscape; others turn to the history of plants and animal life on the site; still others see the vegetal as a repository of historic values, or search for cultural history in order to retrieve ancient beliefs which are still alive in the memory of local people...Many have expressed a concern about issues which are confronted nowadays in different terms by an increasingly environment-concerned populate. Contrary to folklorists or revivalists, none of these landscape creators attempt to create an illusory representation of the past, or a support for nostalgia. They rather aim at building upon an evocation of the past a situation that enable people to move forward, to turn their mind to the future, in full acceptance of the passage of time. This leads to issues of representation of the past course of a landscape thru history, which these artists approach by constructing poetical abstractions of nature. In the words of Bernard Lassus: "An artist cannot produce a literal rendition of the world. As soon as he realizes that he cannot reproduce nature, he has to explore the gap that exists between his work and nature. I think that art is to a large extent concerned with the discovery of the liminal world beyond literal evocation of real objects."

The challenge is huge to ancient civilizations that had a brilliant history, yet slow in its modernization, such as China, Persia and India. Insights of designers and artists from this group equip us with new ideas and perspectives to connect the historical and the contemporary. We have already noted that Lassus sees the historical diversity of cultures as the source of the diversity of landscapes that have left traces on a site, and that he seeks to give a sense of this diversity by bringing to life the different cultural strata which have left some traces and can be brought to the contemporary public's attention. Maya Lin has traveled another path of creation. I would like to turn for a minute to the well-known Vietnam Veterans Memorial of which she was reluctant to speak much in the interview.

At first sight, the two large wall set into the ground may not strike as an expression of diversity. I want, however, to show in which ways it embodies typically Western ideas of diversity, and calls upon Asian cultural attitudes. The memorial was designed as an anti-monument, the opposite of the two nearby great memorials towards which its two v-shaped walls are pointing-the obelisk of the Washington Memorial, and the rotunda of the Lincoln Memorial. It uses typical post-abstract-expressionism artistic gestures to set itself apart from the neo-classical monumental architecture around it, introducing an unexpected heterogeneity in the national mall of Washington DC. Instead of protruding from the earth, and embodying a single individual separated from the mass of common people by a symbol of gigantic architectural proportions, the Vietnam Memorial sinks itself in the earth and celebrates the crowd of veterans who died at war. However it does not look like a tomb at all. It has a pastoral, rather than architectural, quality. It was meant to be accessed through the lawn which slopes gently from the surface of the mall down to the foot of the wall. This lawn itself is an unmistakable reference to the great American lawn, the symbol of the American dream of dwelling in a house set in nature. The American lawn is a cultural symbol of nature, and it is seen by most Americans as a sign of their love for nature. It is a highly reductive symbol of the transcendental encounter between man and nature, as hailed by Ralph Waldo Emerson (1803-1882). Thus in a somewhat paradoxical way the uniformity of the lawn should be seen as a symbol of the diversity of nature, in the context of 20th century America. The Vietnam Memorial was such a success that the crowds walking down the lawn destroyed its perfection, and the problem had to be dealt with by the addition of a path in front of the two black walls on which are inscribed 58195 names of fallen soldiers, irrespective of their gender, race, or ethnic origin. Here again one may see this as a negation, rather than an assertion, of differences. However one should remember that American cemeteries of the two World Wars in Europe are made of pristine fields of anonymous tombs, highlighting the sacredness of the sacrifice of American people, symbolized by the number of dead soldiers. To the contrary, at the Vietnam Memorial each of the dead is identified by name. As visitors walk along the wall they see themselves reflected among the names of thousands of dead, there is an unmistakable kinship between the visit of millions of Americans to the memorial and the descent of Western heroes to the Elysian Fields as described by Homer, Virgil and Dante. Each American visitor repeats the gesture of Ulysses, Aeneas or Dante himself looking for a forefather, a member of his kin, among the dead ghosts who reside in the Elysian field beyond the Styx. The names evoke the bodiless spirit of the dead. Behind the homogeneous representation of each of the soldiers, the name of a family member appearing against the background of their own mirror image amid the crowd of other visitors' mirror images reminds the visitors of their spiritual uniqueness. This presence transforms each visit into a singular experience. It is the experience of an irreducible difference between humans that is brought to light in the ritual of rubbing the name of the deceased, which is encouraged by the park authority and often photographed. Even though Maya Lin did not consciously refer to her Chinese origin, such a connection between the living and the dead calls to mind the Chinese ritual of family commemoration of the ancestor in the tomb-sweeping day, and the traditional art of rubbing.

The Vietnam Veterans Memorial is a work of art that has triggered an even stronger quest for variety. Some veterans felt that this anti-monument did not honor their sacrifice to the nation, and demanded successfully that two groups of sculpture be added to the memorial. They did not understand the diversity that was present in the memorial, or they wanted it to be represented according to post-WWII sculptural tradition. With the additions they obtained, American diversity has been represented according to the American cultural conventions: gender and racial or ethnic differences have been symbolized by groups of soldiers grander than life. This is rather a pale image of the diversity that is already encapsulated in the walls; but one may take it as an unconscious response to such diversity. In summary, the Vietnam Veterans Memorial as a whole introduces heterogeneity in the monumental landscape of the Washington National Mall, it links to recent history and to European cultural history as well as to Chinese traditional practices of commemoration, as well as it makes clear formal references to contemporary art. This is an art that turns history and cultural traditions into sources of new creation. In this light, Maya Lin's call at the end of her interview for Chinese designers to strive for a balance between "what is old and what is new" is indeed a call for creation with cultural and historical sensitivity.

All the designers and artists I interviewed expressed their call to a Chinese creativity for the new landscape. I asked each of these interviewees to give advice to Chinese designers. Out of respect for Chinese culture, none of them has made specific recommendation, but called for their Chinese colleagues to explore a landscape art that delves into aspects of the thousands of years of landscape culture in China, and still addresses contemporary issues. This may seem like an impossible task, and yet Bernard Lassus noted that there was an uncanny kinship between his own strikingly contemporary gardens for the Colas foundation, and the gardens of Suzhou: "They do not look like my work, but there are aspects of the attention for creating an ambiance, for avoiding the literality while remaining evocative, which are absolutely common to these gardens and to my work." These remarks invite Chinese artists to revisit, from a contemporary perspective, poly-sensoriality, ambiance, non-representation, non-action, engagement with nature, fragmentation, discontinuities, anisotropies, relationships to non-humans, and the composite world of Chinese arts (including calligraphy and rockery) involved in the historic transformation of places in nature.

Last but not least, I would like to thank the journal of *LAChina* and China Architecture & Building Press (especially editors Yishuang She and Huaibing Zheng) for their support in making these information available to a broader audience. On a personal ground, it is particularly thrilling for me to have this book published in China, a civilization with arguably the longest continuous landscape culture in the world. A culture that could be more vibrant now in China than ever.

Xin WU
2011

编者按

　　这本访谈录的采访对象有七位，涉及中国、美国、比利时、瑞士等国家的景观设计师。起初我担心它会凌乱，审稿之时，竟然一下被它吸引，一气呵成，连续读完。全书围绕景观设计的一些思想，环环相扣，并无冗笔。我赞叹采访的设计者——吴欣教授对问题的设计，每一个问题都能很好地让被采访者准确谈出其设计的思想精华；我也赞叹被采访大师们的睿智，他们都能简洁地凝聚他们思想的精髓。这对我国的景观设计师来说无疑是雪中送炭，对开拓设计视野会十分有益。故将此书付梓出版，以飨读者。

编　辑

2012.7.20

目录 / CONTENTS

前言——复原景观设计的艺术性
Preface: Restoring Landscape Design as An Art

001 俞孔坚——乡土景观与现代化
Kongjian YU (China) — between Chineseness and vernacular modernity

032 戴安娜·巴摩里——城市景观与自然美
Diana BALMORI (USA) — between urban landscape and natural beauty

064 贝尔纳·拉素斯——视觉艺术与景观的诗意
Bernard LASSUS (France) — between visual art and landscape poetics

096 帕特里夏·约翰逊——公共艺术与环境基础设施
Patricia JOHANSON (USA) — between public art and environment infrastructure

123 埃里克·董特——当代美学与欧洲造园传统
Erik DHONT (Belgium) — between contemporary aesthetics and European garden tradition

156 林璎——艺术创作与形式
Maya LIN (USA) — between art creation and form

180 帕奥罗·伯吉——景观设计与创造性的诠释
Paolo BÜRGI (Switzerland) — between design and creative interpretation

201 附图：被采访人及其设计团队作品展示
Appendix: Interviewees and their design team project exhibit

俞孔坚
——乡土景观与现代化

  俞孔坚是一位中国设计师、作者、教育家和倡导者，致力于在中国建立一个全新的当代景观设计学体系，从实践到教育到理论到公众舆论。他现任北京大学建筑与景观设计学院教授和院长，是北京土人景观与建筑设计研究院（1998）和北大景观建筑研究生院（2001）的创立者。近年来，他演讲著述频繁，土人景观的作品在国内外获得了无数设计奖项。他同时也应邀担任了一些国际评委和哈佛设计研究生院的客座教授。1992年，俞孔坚在北京林业大学任教 5 年以后前往哈佛设计研究生院攻读设计博士学位。毕业后他在美国 SWA 事务所见习了两年。1997年，他回到中国工作并逐渐成为世界范围内当今景观设计的领军人物。不论是以前还是今天，俞孔坚首先始终是植根于中国景观文化，深系土地（表现在他所选择的公司名称和座右铭——"土人"）。正是这种情结使他能够将麦克哈格的环境规划与中国的风水传统相结合，发展出一整套他自己的同时兼顾自然与文化可持续性的"生存的艺术"。作为 20 世纪初已降中国、从未曾间断过的关于现代化／西方化／传统的大讨论的一部分，他对理性规划、城市化妆和文人园林的批判曾经引起了许多争议。俞孔坚探索的足迹为我们提供了许多优秀的佐证——重新省思景观设计学和现代西方学科在非西方文化中的借用。他所取得的成就，不仅对当代中国极有价值，而且在更广泛的意义上丰富了我们关于全球化背景下的景观设计的思考。

吴欣：首先想请问你自身思维的发展历程，你为什么要出国？

俞孔坚：20世纪八九十年代出国是一个风潮，大家都在考虑出国。我出国前已经在北京林业大学任教。我是1980年入学，1987年研究生毕业就留校。当时学校有规定，留校5年才允许出国，我就认认真真教了5年书，各方面也得到了一定的肯定。国内当时的条件确实很艰苦，一家人3代同堂住在一间12m²的房子里。但我还是很积极向上的，所有钱几乎都用来买书，现在的好多书还是那时候买的。小孩儿在旁边睡觉，床头就是我的办公桌，都是书。最多的思考、积累应该是在从研究生到任教的这8年时间，是中国的开放带来的新风。当时积极向上的社会风气，给了我许多营养和进取的动力，那时的一曲《希望的田野》至今仍令我热血沸腾。至于为什么要出国，简单地讲是想呼吸一下新鲜的空气。记得我教书时，在课堂上已经把国际上的一些新东西介绍给了学生，而这些新的东西与传统园林学科所教、所学的都太不一样。我的研究也和当时的园林专业似乎有很多不和谐之处。所以，周边的人包括我的同学在内都感觉我有点怪，不现实。

吴欣：那你当时主要的思考是什么？

俞孔坚：当时的想法非常多，20世纪80年代是中国一个文化高潮，大量的西方哲学，海德格尔、尼采都是那时候涌进来的。各种思潮真是启发了很多东西，对我有影响的包括达尔文主义、进化论美学、弗洛伊德的心理分析、社会生态学，还有地理环境决定论、文化人类学等。这些理论都是把审美与人类进化，文化与景观深深地结合在一起，讨论人的生存经验对人审美的影响，生态环境与社会发展的关系等。在专业上开始接触麦克哈格（Ian McHarg）的《设计结合自然》（Design with Nature），听起来好像离该书的出版已经有近20年的时间了，可那时是中国封闭了30多年后的开放，所有前所未闻的思想和理论突然如浪潮般涌入中国，而我则有幸拥抱了这些浪潮。并开始怀疑我在课堂上学习的东西，特别是对中国传统园林有了批判的意识，对宏观的生态与环境问题有了感觉。对景观生态学、环境伦理以及景观美学之类的东西思考得很多。另外，对中国传统文化包括风水、易经之类也有所钻研,甚至还发表了不少与专业无关的论文,包括提出易经发源地及其景观的探索性文章。你可以想象我对景观的理解跨度是很大的。

吴欣:这些想法你在出国之前就有了,对吗?

俞孔坚:对,我走之前麦克哈格的书就已经读过了,还包括《景观生态学》(*Landscape Ecology*),最早是1986年出版的,这本书在翻译出版之前就已被我当做给学生上课的教材,书的作者佛尔曼(Richard Forman)后来就成为我博士指导委员会中的一个导师。当时我在北林开了一门关于景观的文化、生态与感知的课,课程名称也就是后来我第一本文集的名字《景观:文化、生态与感知》(1998)。

我最早的研究及毕业论文是风景评价,涉及人的美感是怎么来的,人是怎样审美风景的,又试图通过科学、客观的方法来测量。当时对我影响较大的就是刚才讲的哲学、心理学方面的书,包括《信息论美学》,主要讲视觉信息。还有一本很重要的书叫《景观体验》(*The Experience of Landscape*,作者 Jay Appleton),这本书就是从进化论的角度谈论美学问题,它提出了 Prospect-Refuge(瞭望-庇护)理论,就是看与被看的空间关系。还有风景评价(Landscape Assessment)方面的书,讲怎么评价景观和进行景观感知的研究。我们以前学的都是谈形式美的评价,用传统那套线条、色彩分析的方法来评价。接触这些以后,想法突然被改变了,不光是形体、色彩,实际上人的更深层次体验是跟人的进化紧密联系在一起的;这就是我为什么一直研究风水,最后提出中国人的理想景观模式,从人类进化和文化发展的地理环境来理解中国人的景观理想。

接受了这些新理论后,我对园林有了不同的看法,对中国传统园林的理解有别于教科书上写的那种线条的分析、色彩的分析、形式美的分析、意境的分析。所以在1987年我就写了一篇文章,叫《中国人的理想环境模式及其生态史观》,分析了中国人为什么会有这种理想的景观模式,分析了中国人为什么喜欢被庇护、为什么喜欢四周围合、为什么需要空间的小中见大。我总结了一个葫芦模式，它是中国人理想景观的普遍模式。原因是人的进化过程，特别是中国文化的演进中对安全感和庇护环境的偏好，并影响到了景观美学。

在我出国之前的 1992 年就完成了一本书，叫《理想景观探源：风水的文化意义》，用生物与文化的基因解释中国的风水，但由于当时出版社较忌讳"风水"二字，所以，直到回国后的 1998 年才在台湾先出版，然后再在商务印书馆出版。关于理想景观的生物与文化基因，必须分两个层面来解释。首先是生物基因，凡是人都会选择好风水这样的环境，这是东西方共有的，但为什么会有不同的景观偏好，那就是中国地理环境决定的，因为文化在进化过程中有不一样的地理环境经验并产生与之相适应的理想模式。诞生在地中海岛屿的西方文化是一种贸易型的文化，是一种扩张型的文化，这就形成了西方理想景观模式的特点，希望控制制高点，喜欢视野开阔。中国文化的进化是以农耕为主导的，靠捍卫农业生产的领地来维护社会发展；所以中国人喜欢盆地，喜欢"左青龙、右白虎、前朱雀、后玄武"的围合结构。中华民族发展出了封闭式的、盒子套盒子的景观偏好，而西方会发展出开阔的、点到点的、放射状的景观偏好。西方人为什么喜欢在尽端路头盖房子？西方轴线为什么很直？它跟故宫、圆明园为什么会不一样？这都是东西方的文化差异导致的。同时，生物基因为了生存，发展出两条战略，就是进攻与捍卫，动物、人都是这样。要么你的进攻能力很强，要么你的藏匿能力很强。而中国的景观偏好，特别是园林空间偏向于藏匿，西方则偏向于进攻。之前关于中国的园林是没有这么分析的。

国外的关于景观规划设计的理论和思想，在很大程度上改变了我的一些看法，这其中还包括哈佛大学设计学院教授斯坦尼兹（Carl Steinitz）的系统规划方法论。我是在出国之前，斯坦尼兹到北京林大做系列讲座时认识他的。当时因为与中国的学科认识差距太大了，能够听懂和理解他所讲的人并不多，我是其中之一。这很大程度上得益于北林作为林业部图书进口指定单位的优势，我那时候已可以直接接触到好多英文文献，比如 *Landscape Architecture* 和 *Landscape and Urban Planning* 杂志。这样，我在国内时就得以跟踪国际上景观设计学科的最新发展，申请博士学习是有目的地要出国学这些新的东西。

吴欣：照你刚才的逻辑，中国园林是最符合中国文化和中国人的本色，那你为什么要批判它？

俞孔坚：这是两个概念。中国园林的空间趣味反映了中国人的景观理想，是个唯美的角度，但它必须放在社会和环境伦理下来讨论。同时，其形式、内容和材料及技术都与当下时代产生了距离，同样的理想空间完全可以有不同的形式、材料来体现。综合考虑社会和环境伦理以及美学问题，我主张用真、善、美来综合评价设计景观。在这样的标准下来认识中国传统园林，首先它经不起"真"的考验，假山假水，里面的观赏植被也缺乏真实性；其次，它经不起"善"的检验，其铺张浪费之极，多少帝王贵族为造园林，不惜倾家、倾国，弃人民于水深火热，而独自享用；唯有"美"字，仁者见仁，最说不清，所以不必评价。当然，我用这些标准来评价中国园林对于一个文化遗产来说，是不公平的，因为它的产生时代与我们今天不同。之所以我又要如此批判，是因为有许多人要发扬光大这种应该反思的传统，让它复活，误认为中国现代还应该造这些东西，那我就不得不把它对当代中国的危害性揭露出来。让人们认识到"国粹"的真实面目。

传统的文化，是有历史局限性的。为了保护历史的含义，我们只要将遗产保护下来就行了，就已经标榜了历史的阶段和文化身份了。历史悠久并不能代表有多么先进，如果历史很悠久，但社会或学科总在原地踏步，徘徊不前，那么这样的历史是垃圾，没有意义，不值得自豪。历史性或历史意义在于它的变革、创新及其对社会发展的价值，而不是它持续多久。新的时代需要有新的景观和建筑，要创造新的文化，就完全可以摆脱所谓的旧传统，当然如果旧传统有用，自然而然可以保留下来，如果没用、不适应当代的要求，那只能作为遗产来对待，我的价值观就是这样。你要创造这个新文化，不一定非得再做一个老帽子上去或者再穿一套老衣服上去，新酒没有必要装在旧瓶子中。我的这些观点，在西方学者看来是不言而喻的，因为他们经历过文艺复兴。在中国我的这些观点要被大家所接受，却显得艰难，因为现代主义运动在中国没有彻底，在建筑和园林领域甚至没有沾到边。传统园林（包括西方经典的园林）首先不符合当代的环境伦理、不符合可持续原则、不能满足当代人的生活方式，也不能彰显当代人的文化身份。中国的园林主要是封建士大夫文化和生活方式的反映，根本不能解决当代中国所面临的环境和社会问题。应对当代所面临的环境、能源、人口的问题，必须有适合当代的一种方式。从这个角度看中国古典园林，第一，它造价昂贵；第二，它只能是少数人的天堂；第三，它缺乏地域性；第四，它缺乏环境伦理和生态意识。园林可以造得很诗意，但是它违背了当代的环境伦理和价值观。我们必须有更高的价值标准来衡量什么是美的，什么是不美的。需要新的美学，这种美学我把它叫"足下之美"或"低碳美学"。

吴欣：那么你找到"新乡土"了，对吗？

俞孔坚：我认为是，适应于当代中国，解决当代中国环境问题并服务于当代中国人的"新乡土"。

吴欣：这个"新乡土"是不是就等于城市风水？

俞孔坚：不能完全这么说，实际上风水和园林是两套体系。中国园林里头从来不用风水原理，风水先生跟中国园林没有关系。风水是用在老百姓看墓和选宅基地上的，是一个乡土的东西、实用的"生存艺术"，当然也被帝王所御用。所谓乡土是日常的、活的、因有用而衍生的。中国园林反映的是一个士大夫的消遣的艺术，这是两个层次的文化。所以说我最后认识到，风水实际上是一个关于人对生存的环境适应性的文化，而园林纯粹是一种唯美的艺术。

虽然说风水实际上是一个前科学时代的东西，但是我觉得景观设计应该至少与风水有同样的理想，就是人跟自然的联系、人跟土地的和谐。风水是乡土的一部分，而乡土就是特定人群对环境的一种适应方式，它是使用者创造出来的，有功能、满足日常生活需要的。我们的环境建设，包括城市园林建设，必须考虑中国的环境和资源危机、粮食安全、全球的气候变暖等问题，这是定义新乡土很重要的方面。旧乡土是中国古代广大人民自发的、为适应环境、实现生存和生活的一种方式。新乡土则是适应当代环境挑战而形成的,因必须解决脆弱生态系统下这么多人的生存问题而产生。现代中国城市的生态环境条件显然已无法承载古代士大夫那样的生活模式。

新乡土是一种适应当地气候条件的、满足当地人生存并解决环境问题的最节约的方式。因为乡土最节约,最简而易行,所以形成了独特性。乡土一定是适合当地的,土生土长、简单而经济的。只有乡土才最节约,耗费自然资源最少。所以,我们必须放在当代环境下来认识新乡土。

还有一个层面的认识,就是旧乡土没有被上升到上层文化,它是没有被高雅化的。中国从乡土文化进入到上层文化,它是随着科举制度、城市化,由土变洋的过程,所以中国当代城市化出现了大量的非乡土东西,比如说巴洛克风格的景观。西化和现代化并不是一个概念,现代主义并不就是西方的,现代主义完全可以是本土的,完全可以是既现代又是中国的。实际上这是后现代的概念,强调地域性。现在有人说我做的景观设计是西方的。他们实际上混淆了现代和西方这两个概念。现代主义和地域性这两个概念并不矛盾,现代的东西可以是乡土的。另一方面,如果把传统的东西跟地域性的东西混为一谈,就把两个概念都简单化了。把所有现代的东西都认为不是中国的东西,这是目前一些人头脑中存在的最大误区,也是最大的心理障碍。老套到传统的中国形式里是不行的。现代主义设计是不是可以产生中国的东西?完全可以。

实际上我们就是在探讨一个新而中的问题,不要刻意到中国传统里去寻找所谓中国,只要你解决了中国的问题,只要你符合中国人日常的需要,用了当代中国技术所能做的东西,适应当代中国特定地区的环境,又是中国人去做的,怎么不是中国的?这完全是中国的。"红飘带"就是中国的,它解决了新乡土这个问题了。岐江公园也解决这个问题了。

吴欣:"红飘带"有现代主义设计的影子。我知道你曾经在彼得·沃克(Peter Walker)和 Sasaki 的事务所工作过,你对彼得·沃克的设计怎么评价?

俞孔坚:彼得·沃克是现代主义景观设计的核心代表,他把现代建筑的简约风格引入到了景观,将现代艺术引入景观设计,他这种Minimalism(极简主义)体现在形式上就是整形的、简洁的,他的所有东西都是几何性的,这个几何性跟现代建筑是完全融为一体的,这是他的成就。他强调了现代主义的形式,把建筑和艺术中的Minimalism变成了现代形式主义的园林。他对生态并不感兴趣,如他很少用野草或不简洁的东西,他的绿篱是年年需要修剪的,他摆的那石头是非常精心设计的,动一下其中的一块,形式的美就要被破坏。

彼得·沃克对我的影响,就是他的形式美,简洁、简约的美。但在我这里,你会发现首先我把材料换掉了,内容也与他的设计相距很远,生态性和景观的生态系统服务功能是我设计的主题。但我强调生态性并不是回到自然主义和环境主义,而是把生态性、生产性与审美,把环境伦理和艺术性结合在一起。沈阳建筑大学校园用的是稻子,有人说我的形式和他在德国凯宾斯基饭店做的那个景观形式一样,但其实完全不一样。沈阳建筑大学校园是生产性的、包括雨水的收集利用、粮食的生产和可持续理念的强调,它有乡土的内容、生产的功能和当代的生态服务价值。在内容上我更像是回到麦克哈格的理论那里了,但我又不同于麦克哈格,因为我重新把艺术带入到环境,环境主义跟现代的艺术在我的设计中都得到了体现。对国际上的环境保护主义的设计,我也试图有所创新并显示自己的特性。如岐江公园跟理查德·哈格的西雅图煤气厂改造的公园不同,和鲁尔钢铁公园也不同。西方的这两个案例中并没有用太多的新的设计,更多的是保留旧迹和土方,解决了污染废弃物的堆放和旧机器的展示,更多地强调了环境主义的内容。而岐江公园不光是生态的恢复和环境主义理想,而且很强调新设计和新体验、铁路、路网的设计,红盒子完全是新的设计,艺术融进去了,同时环境和生态也进来了,包括野草的使用,有了些后现代的味道。永宁公园也不光解决了防洪的问题,还有艺术的设计。所以艺术设计是受到彼得·沃克等现代简约大师的影响,但内容上又受到当代环境主义和生态伦理的影响。简约的形式还在,当代的艺术的东西还在,但都是在生态的背景下存在的。我的设计是生态的简约主义(Ecological Minimalism)。种一片野草,恢复个湿地,我觉得是好的,但那仅仅是个生态原则,一定要有艺术性,一定要有当代人的意识在里头,所以我强调生态与人文同时存在。把生态、人文和艺术结合在一起，就可以形成真正的当代的景观，中国的新乡土也在这里。

这两者结合也可以产生中国自己的特色。你如果完全按照生态去做，那中国的景观也没有什么文化特色，因为那是自然过程，这就变成了彼得·沃克所说的"看不见的花园"，没有文化意义在这里。必须两者结合。没有艺术的设计，人就不存在了，人的意义就没有了，这个专业也不存在了。彼得·沃克有本书叫 Invisible Garden（《看不见的花园》）。他实际上批判了唯生态主义，开始提倡景观设计学的设计含义，出现了现代艺术基础上的设计。那么我觉得我们是在这个基础之上又有一个否定之否定，重新回到生态，但我们已经不是原始意义上环保主义的生态概念了，而是艺术的生态。

我们最近的作品天津桥园就体现了生态的简约主义，在一块城市废弃地上，挖了一个个坑，像一个个水泡，收集雨水，植被群落围绕这些坑进行自然演替，然后把人体验之木平台放到里面去，泡泡的生态美就显现出来了，也可以说这是艺术，生态的内容恰恰也进去了。不同深度的坑，水分就不一样，盐碱度就都不一样了，就导致不同群落生长，这完全是生态的内容。但这个坑的布置，坑的深浅完全是人为的。这和"红飘带"的方法正好相反，桥园是在艺术里加生态，红飘带则是在生态环境里放了艺术。这两个都可以起到很好的效果。

中国现在最大的问题就是环境问题，解决环境问题最关键的策略就是重建人与土地，人与水的联系。城市的景观设计应该重建人跟土地的关系，让人能看到雨水是怎么利用起来，变成地下水的一部分，就是自然过程的可视化。生态基础设施实际上就是重建人跟土地的联系，重新让人跟大地发生联系，重新通过生态网络恢复城市中的雨水过程，复活当地的植被、文化遗产及人的活动的联系……

实际上欧美也在反思，建筑学走向了绿色建筑，这时候就可以看到建筑学跟景观设计学是平行发展的。库哈斯（Rem Koolhaas）的那套怪诞的建筑和结构形式不能继续发展下去，我们最终还是走向生态的、平常的建筑。景观也是，你走形式是不行的，最终要走向生态的景观。在更大尺度上就是绿色的城市、绿色的国土，那就是我提出的景观安全格局。

吴欣：这些要进入到规划的层面，能讲讲你的"反规划"概念吗？它是不是还是城市规划，但只是不同方式？

俞孔坚："反规划"是规划的不同方式，是对传统的规划方法论的反思和改进。它是建立在中国目前所遇到问题的认识基础之上，以及对现行中国规划管理和规划方法论的局限性的认识基础之上。"反规划"可能在欧洲就没必要提了。因为，欧洲大面积的规划已经做完，或者说大面积的城市建设已经停止了，不需要这么去做。"反规划"更是针对中国如此快速的城市化背景，你没做完建设规划，土地就已经城市化了，开发商早已在城市规划区范围外圈地搞建设了，而我们的规划部门还在做城市建设区的规划，你的规划设计的理想模式再好，它没法按你的规划去做。与其这样，不如先做一个能覆盖全国土的、不建设的规划，不如先做一个避免犯错误的规划，一个1000年不变的规划。湿地、河流、文化遗产在建设过程中往往被忽视，所以首先把它们作为不准建设区域，作一个法定的规划，让它们永远可以存在，那你的建筑盖得再丑也就无害大局，有人会去拆，10年、20年以后拆掉。但是河流廊道和生态基础设施留下来了，湿地永远留下来了，这就是"反规划"。它是保证不犯大错误的规划，不至于犯伤天害理的建设的规划，在开发商无孔不入的今天，没有比这个更紧迫的了。

吴欣：你怎么保证当地政府能跟踪执行？

俞孔坚：通过各种途径呼吁并强调它。这也得益于我有许多机会能给市长们讲课，各地政府也愿意请我给他们讲课，效果是很显著的。在这些讲课中，我强调它的重要意义，强调"反规划"的成果应该立法。它可以融在城市规划中去，但城市规划必须遵循着这个来做，你做城市总体规划之前，必须做"反规划"。另外，"反规划"并不是限制你去发展，"反规划"只是告诉你，哪些地方你不准做，政府就知道当项目进来时，可以安排在别的地方。我的底线在这儿，这个底线谁都可以接受。我只保留下30%、甚至更少的地方，剩下70%的地方都可以搞建设,也就是说给了他一个更大、更宽的范围去做建设规划。"反规划"是在中国目前城市快速发展、环境资源很脆弱、开发压力很大、开发商无孔不入的背景下提出来的。但同时具有普遍的意义,因为传统规划往往做一个完美的设想,而城市扩张的时候总有许多不可预测的因数。所以保护比预测更有可行性。

吴欣:土人提出"足下文化与野草之美",能否谈谈你对现在时髦的农业旅游的看法?

俞孔坚:土人提出的"足下文化与野草之美",是试图通过我们的作品建立一种新美学,它基于当代环境与生态伦理而产生。而农业旅游是一种对生存艺术的体验,与传统的名山大川的猎奇式观光旅游完全不同,也反映了社会发展对新美学和新的景观体验的新趋势。农业旅游让城市人重新回到土地,体验到土地的丰产,土地的真善美,当然,它还是一种新经济。它是城市化的社会对土地的回望和依恋。这也意味着农业景观和生产性景观将成为未来景观设计的一个重要领域。景观设计从收珍猎奇,走向日常与寻常。

吴欣:能否谈谈其他设计师应如何受益于土人经验,在获取灵感的同时又避免落入形式化?

俞孔坚:我想土人的经验根本体现在它的名字中。他对土地的热爱,对协调人与土地关系的热衷,对乡土的情怀。大家看土人的作品似乎有一种统一的风格,实际上土人并没有刻意去追求什么风格和个性,如果说土人的作品有风格、有个性,那是因为土人的设计显现了项目所在地的特色和个性。没有两块土地是一样的,正如没有两个人是一样的。景观设计的灵感来源是土地本身,如果遵循这一原则,就不必担心落入形式化了。对待共同的问题会有不同或相同的解决途径,但如果把问题放到具体的地域环境中去考虑,那么,相同问题的最好解决对策也会是各不相同的。当然,要做到这一点,敬业精神变得非常重要。所以,土人的经验不仅在认识论和方法论上,更重要的是敬业精神和职业态度上。这方面我们也并不是完美无缺的,遗憾也会时常发生,但这是我们自己都常常引以为戒的。

吴欣：最近结束的哥本哈根联合国气候变化大会使许多人开始认识到环境问题的复杂性。气候的问题不是仅仅科学或立法能解决的，同时关系到一系列论题，如能源与消费、历史与文化、经济与发展、贫困与人口等。很多设计界的有识之士还将矛头指向过量建筑和建筑方式，认为人类必须重新思考什么是设计和为什么而设计的问题，如果不从根本上改变现有对"建设"的认识，任何建设都是对自然环境的破坏。景观设计业无论有多"绿"，最终还是人为的建设。你对此有何见解？

俞孔坚：完全同意你的看法。首先，我们建了许多不需要建的东西，无用的东西，消耗了大量的自然资产。世界上有，中国也有不少空城。第二，是在不该建的地方建了需要或不需要的东西，典型的是城市建设侵占河漫滩、湿地等生态敏感地带，结果使建设成为破坏性的建设，因为这种城市建设毁掉其赖以生存的生态系统基础，报应是无需长久就可以看到的；第三，建了一些需要高成本维护的建筑，成就成为负担。在景观上，问题是一样的。

关于如何减少碳排放，实现低碳生活和低碳城市问题，最近已成为人类头号难题，几无良策。阻止温室效应比打倒希特勒还要困难许多。因为这涉及改变人类存在的最基本的生物基因和最基本的信仰问题。人类天生似乎多挥霍的本能，而少节制的基因。在一个开放而无限的资源环境中，人类的无度挥霍欲是人类不断进化和发展的动力。但人们突然发现，地球只有一个。于是，无穷的挥霍欲和有限的环境承载力之间便出现了难以调和的矛盾。人们幻想通过科学技术提高满足欲望的效率，来减少碳排放，比如飞机如何更高效，汽车的排放如何更小，房子如何更节能等。事实是，科技本身不能解决地球升温的问题。所以只要人类还有扩展欲望的空间，任何科技或节能、节约的技术，都将无济于事，而只能饮鸩止渴。正如网速的提高不能减少我们办公的时间、马路的宽敞带来的是更多的汽车一样。

关键在人的欲望。如果人类的欲望从一个无限量变成为有限值，拯救地球就有希望。显然大多数人不能像佛教徒苦行僧，但我们应该有共同的信念——与景观环境直接相关的是美学。先贤蔡元培早已就美学代替宗教有过论述。如果我们的人民能有高尚的审美观，比如认识到装饰性的绿化实际上很丑陋，华奢的建筑实际上是怪诞而非美丽，走路和骑自行车比开汽车更有风度，小区里种上蔬菜和玉米实际上很美……那么，我们离低碳生活实际上已经很近了，这个地球也会更持久些。

总之，我们必须认识到人类的挥霍本能和无度的欲望是低碳生活的头号敌人，而科学和所谓的高效并不是良方。我的解决办法是全力提倡新美学和对土地宗教式的爱护。

吴欣：你的批判态度与20世纪初五四新文化运动一致。蔡元培先生自己可是留过洋的前清举人。我注意到你对古文颇有修养，请你谈谈对中国文化传统的看法。

俞孔坚：这是个大话题，一言难尽。我尽管在美国5年，我的主要教育和思考都是来自于中国，我从来没有一刻忘记过自己是中国农民的儿子。对待中国传统文化，我所持的观点受胡适和鲁迅那一代新文化先驱的强烈影响。我认为：总体上讲，中国文化应该分为两个层次来理解：一种是生机勃勃的下层乡土文化，它丰富多样，一直延续至今，它对中华民族的生存和发展发挥过、并将继续发挥积极的作用。与我们专业有关的这种乡土文化包括：如何对待土地，如何选址和造地，如何理水，如何种植，如何盖房子等。它在不同的地域环境中，在极有限的土地和资源条件下，解决了众多人口的生存和繁衍，持续至今。另一种传统文化是上层贵族和士大夫的文化，它与生存问题无关，确切地说是消费和挥霍的文化。我们的世代文化传承往往注重上层文化，而忽视下层文化。在当代环境危机面前，我们应该继承和发扬的是生存的艺术，是乡土中国的活的文化，而非历代帝王和士大夫的那种死文化。中国当代许多人高谈阔论如何继承和发扬传统文化，只可惜，多是"不问苍生问鬼神"，舍本求末，把一些腐朽的文化遗产当做"国粹"，而对生存的艺术茫然无知。

吴欣：你如何看待美育？

俞孔坚：审美教育是教育中最基本、最基础的。蔡元培说过美育在中国应该代替宗教，或者说像宗教那样重要。我的野草之美就是想发展一种新的美学观，这方面的文章"足下的文化与野草之美"已经收入到中学的教科书里了。实际上，这是用新的审美观，给人们一个冲击、一个完全不同的思考。在淮北的一个中学，老师给他们布置了一个作业，围绕我的文章写心得，就让他们观察本地的、乡土的美。心得写出来都非常感人。他们看到的当地野草是美的、当地鱼塘是美的、当地稻田是美的，这就是新美学与新乡土观的意义。这其实跟新文化、跟文学革命是有关系的。中国需要这样的一个新的审美启蒙运动。这个审美启蒙运动是五四新文化运动的延续，是美育的现代主义运动，是一个民族要进步的最基本的东西，但让更多人接受它还不是那么容易。

景观设计是一个社会的行为。我们必须同时是教育决策者、发展商、设计师、工程师和使用者；这个学科必须重新定义、重新评价、重新给予标准，一个连基本的价值观都是错误的学科体系是不可能创造出好的作品来的。所以这两年我们工作中很重要的一个方面就是重新定义景观设计学，认为是一门源于大禹治水，源于 5000 年灌溉、造田、种庄稼等中国乡土经验的生存艺术。我把景观设计学定义为"生存的艺术"（The Art of Survival），这可以说是麦克哈格环境设计思想的延续，但更是对中华大地的自然史和农耕史的思考的结果，是对祖先经验和当今挑战的回应。我今天的思想很多受益于回国以来在中国从事的景观教育和实践，但这些认识是有全球意义的。

吴欣：希望你能教出越来越多有想法的好学生。你是北大刚刚创办的建筑和景观设计学院院长，能否谈谈对中国景观设计教育的想法？20 世纪以来世界景观设计领域日新月异，全球化的弊病促使越来越多的个人和国家开始反省景观的文化意义。老一套的现代主义设计观和技术至上论已经过时，历史文脉和地域特色日益得到提倡。你对在外留学的中国和其他亚洲国家学生们有什么特别的建议？

俞孔坚：首先，我这里谈的景观设计学科与传统的风景园林学科是两回事。中国的园林有悠久的历史。但中国的景观设计学科才刚刚开始，需要开拓和完善，体系需要建立，教师人才奇缺，但前景无限。这个学科必须在关于自然系统的学科、社会系统的学科、历史文化系统的学科以及工程技术系统的学科基础上来建立。土地是景观设计学研究的对象，人与土地的和谐是学科的目标，培养具有当代土地伦理，掌握现代科学技术，能通过规划设计手段来协调人与自然关系的人才是教育的目标。土地与环境伦理基础之上的新美学是这个学科发展的重要根基。生存的艺术是这个学科的基本定位。而发展这样的一个学科，需要几代人的不懈努力。目前来看，社会上对该学科教育的热情已经起来，过去10年的时间为我们未来的发展在荆棘中开辟了一条光明之路；未来10年时间将会有正本清源的工作要做。学科的发展和人才的培养一样，都需要有一个周期。我认为，再过10年，中国的景观设计学教育将会出现一个真正繁荣健康的局面。

关于对在海外留学的中国留学生的建议，首要的是胸怀祖国和胸怀世界，这不是套话，而是一种学习的目的和态度。明确学习的目的是改变这个世界同时改变自己。最重要的是应该认识到中国当前面临的巨大的生存挑战，从环境到文化，应该最有效地利用你宝贵的学习机会，把精力集中到有意义的事情上，不能浪费时间。凡有所学终有所用，不学则无为。

Kongjian YU
—between Chineseness and vernacular modernity

A Chinese designer, writer, educator and advocate, YU Kongjian has been endeavoring to establish a whole new system contemporary landscape design in China, from its practice, education and theory, to its public reception. He is a professor and Dean of the School of Architecture and Landscape Design in Beijing University, and the founder of Beijing Turenscape (1998) and the Graduate School of Landscape Architecture (Beijing University, 2003). In recent years, he has lectured worldwide and the design of his firm won national and international recognition. He also serves as jury in international committees and a visiting professor at Harvard School of Design (GSD). In 1992, he embarked on a doctoral research at Harvard GSD after having taught 5 years in Beijing Forestry University. Following the completion of the study, he practiced at Sasaki and Walk Associates (SWA) for two years. In 1997, he returned to China and has gradually become a leading figure in contemporary landscape design worldwide. Then and now, YU Kongjian is first and foremost a person deeply immersed in Chinese landscape culture, a man firmly tied to the land (as indicated by the name and mission statement of his firm-Turenscape ["Earthman-scape" in Chinese]). It is this aspect that enabled him to bridge McHargian environmental planning and the time-honored Chinese feng-shui tradition, and to develop a body of thinking that concerns both the sustainability of nature and culture, which he termed as "the art of survival". Over the years, his criticism about rational planning, urban beautification and literati gardens has raised much controversy, as part of the unsettling Chinese debate upon modernization / westernization /tradition since the dawn of the 20th century. The trajectory of his exploration provides excellent examples for the reexamination of landscape architecture, as well as the adaptation of modern Western discipline in non-Western cultures. His work not only is proven to be extremely valuable to contemporary China, but also enriches, in a broader sense, the thinking about designed landscapes in an age of globalization.

Xin WU (WU hereafter): First, I would like to learn your development. Why did you choose to study abroad?

Kongjian YU(YU hereafter): The 1980s and 1990s were the peak of study abroad in China; almost everyone thought about it. Before going abroad, I was already a lecturer in Beijing Forest University. I was admitted as an undergraduate in 1980, graduated with master's degree in 1987 and remained in the university to teach. The university rule then was that one could only go abroad after working for 5 years; so I did and my work was appreciated. The living condition in China then was tough; 3 generations of my family live in a room of 12 m^2. But I was optimistic and spent almost all my money on books; many books I have now were bought then. My desk was always piled with books, just beside the bed where my child sleeping. A lot of thinking and accumulation of knowledge were carried out during those 8 years of graduate study and teaching. It was the new era of reform and opening in China. The positive and energetic social environment nourished and inspired me. Even now, I would be emotional when hearing the then-popular song with the title "The Land of Hope". As for why to go abroad, to put it simply, it was longing for some fresh air. I remember trying to introduce new information from abroad to my students in teaching. These new information were not quite the same to what was taught and learnt traditionally in the discipline in China then. My research also did not follow the established pattern in the discipline. So, many people around me, including my classmates, felt that I was strange, unrealistic.

WU: What did you mainly think about then?

YU: A great deal! The 1980s was a heyday of new culture in China. A great number of Western philosophies, such as Heidegger and Nietzsche, were introduced, and various thoughts inspired a lot of new things. For me, there were Darwinism, Evolutionary Aesthetics, Freudian Psychology, Social Ecology, Environmental Determinism, and Anthropology, etc. All these theories incorporate aesthetics and evolution, culture and nature, to discuss connections between survival and aesthetics, ecological condition and social development, etc. Regarding to the discipline, I read *Design with Nature* by Ian McHarg—it was already some 20 years after the publication of the book, China just opened up after 30 years of seclusion. Unprecedented amount of information arrived like tsunami, and I was luck enough to have the opportunity to be immersed in all these. I started to question what was taught in class, especially being critical about classical Chinese garden, and being intrigued by large scale ecological and environmental issues. I thought a lot about landscape ecology, environmental ethics and aesthetics. On the other hand, I also researched traditional Chinese theories such as Fengshui and *The Book of Change*, and published several academic papers, including a challenging study about the place of origin of *The Book of Change* and its landscape. You can imagine that my understanding of landscape is very board.

WU: Am I right that you have had all these ideas before going abroad?

YU: Yes. I read McHarg long before arriving the US. Also *Landscape Ecology* published in 1986, which I had used it as textbook before it was translated into Chinese. The author of the book Richard Forman later became a member of my doctoral committee. A course I taught at Beijing Forest University was about the culture, ecology and perception of landscape; it later became the title of my first anthology—Landscape: Culture, Ecology and Perception (1998).

My initial research and thesis was on landscape evaluation, involving what are the source of human aesthetics, how people appreciate landscape and how we can measure it scientifically and objectively. Some of the most influential books to me then were the aforementioned books on philosophy and psychology, including *Information Aesthetics* about visualization. Another important book to me is *The Experience of Landscape* by Jay Appleton, which addresses aesthetics from the view point of evolution and coins the theory of "prospect-refuge" through the spatial relationship of seeing/being seen. Also were books on *Landscape Assessment*, about landscape evaluation and perception. Most things I learnt before approached the issue of beauty from formal perspective—line, color, etc. After being in touch with these new ideas, my thinking about landscape started to change, realizing that it was connected to not only formal elements but also deeper experiences related to human evolution. This is why I have always been researching on feng-shui, in order to explore the ideal landscape pattern of Chinese people and to understand Chinese ideal of the perfect landscape through human evolution and cultural geography.

After accessing to such new concepts, I started to develop different opinions about

Yuan-lin, started to develop understanding of traditional Chinese garden in a way that varied from the common presentation in textbooks—form, line, color, atmosphere, etc. So, in 1987 I wrote an article titled *"Ideal Chinese Environment Model and Its Eco-historical Origin"* to analyze why the Chinese people prefer such ideal environment, why we love to feel protection, why we like encircled sites, and why spatially we need to see the big from the small. I summarized a gourd model, which demonstrates the ideal Chinese landscape setting. The reason is the evolution, especially the Chinese culture in its craving for an environment of safety and refuge which in turn influences landscape aesthetics.

Before going abroad in 1992, I had finished a manuscript titled *The Ideal Landscapes: The Meanings of Feng-shui*, using bio-cultural elements to explain Chinese fengshui theory. But because many presses at the time had hesitation about feng-shui, it was only published in 1998 after my return to China, first in Taiwan then by The Commercial Press in China. The bio-cultural root of ideal landscapes has to be explained in two levels. First, its biological root—all human beings choose environment with good landscape, which is a shared aspect of the East and the West. As why the Chinese have different approach, that is decided by the geology and landscape of China, since different cultures reach different ideal landscape patterns that echo to their experience of the environment. The Mediterranean culture was based on trade, it was expansive. Thus an ideal landscape pattern that is in favor of vantage points and open vistas. The Chinese culture was cultivation-based, the development of its society depended on the defence of agricultural territory. Thus in favor of basins, with encircled structure of "Left Green Dragon, Right White Tiger, Front Red Phoenix, Back Black Warrior". The Chinese people developed a taste of closed, box-within-box landscape, while the Western taste turned to open, point-to-point and radiant views. Why the Westerners prefer to build at the end of the road (a mistake from fengshui point of view)? Why axes in the West are straight? Why are they different from that of Forbidden Palace and Yuan-ming Yuan? All these are cultural differences between the East and the West. Meanwhile, the biological instinct gave rise to two survival strategies—attack and defence, which are the same for both human beings and animals. Each species either excels in attacking, or in hiding. The Chinese landscape preference intended to retreat, especially Yuan-lin; while the Western one intended to advance.

There was no such analysis of Chinese Yuan-lin like this before.

My ideas were further changed by landscape planning and design theories available abroad, including the systematic planning methodology of Professor Carl Steinitz at Harvard GSD. I first made acquaintances with Professor Steinitz when he gave a serial lecture at Beijing Forest University. At that time, due to the huge back lag in this field in China, only a few people really understood his model; I was one among these few. I greatly benefited from the university's being the official recipient of imported books, a privilege assigned by the Chinese Ministry of Forestry. I already had direct access to many English publications, such as *the journals of Landscape Architecture* and *Landscape & Urban Planning*. So, I already followed the newest developments in the field of landscape in China. When applying for doctoral studies, I purposefully planned to learn these new things.

WU: According to your logic just now, Chinese yuan-lin fits perfectly the culture and preference of Chinese people. Why do you criticize it then?

YU: There are two different concepts. The spatial organization of yuan-lin reflects the ideal landscape of the Chinese; this is an aesthetical dimension which has to be examined further under social- and environmental-ethics. In the mean time, when the form, content, material and technique all changed in the contemporary, the same ideal of landscape may be expressed through new forms, materials and experiences. I believe in truthfulness, goodness and beauty as the united standard to assess designed landscapes, that is to consider aesthetics in the complex of social- and environmental-ethics. To rethink about traditional Chinese scholar's garden under such criteria, firstly it is hardly "truthful"—with artificial rockery and waters, plus ornamental plants; secondly, it is hardly "good"—numerous emperors wastefully built exclusive luxurious gardens that brought disaster to the country and people; as whether it is "beautiful", that is still open to discussion. Of course, from cultural heritage perspective, it is unfair to evaluate Chinese Yuan-lin in such a way, since it was the product of a totally different era in history. Why do I criticize it so? It is in fact points toward the revisionist tendency that we should still only build such gardens in contemporary China. I feel I have to reveal yuan-lin's short-coming in order to count-balance essentialism.

Traditional culture has its own limits. It is necessary to preserve culture heritage in order to mark historical periods and cultural identity. But long history does not equal to advancement. History would lose its meaning and pride if our society and science remain stagnant. To me, the significance of history lays in its value of stimulating reform, innovation and development, not its length. A new era calls for new landscape and architecture, calls for the creation of a new culture; it is totally possible to escape the fetter of the old. When tradition is alive, it certainly should be continued. When it is no more alive, unfitting to the modern needs, then it can only be seen as heritage. So are my values. To create a new culture, it is not necessary to cover it with old appearances—new wine does not need to be put into old bottles. These opinions seem natural in the West due to its experience with the Renaissance; but very difficult to be accepted in China since modern movement is incomplete in China and almost none in the fields of architecture and landscape. Traditional gardens (including historical gardens in the West) do not match contemporary environmental-ethics and the principle of sustainability. They can neither satisfy the new lifestyles, nor promote the new cultural identities. Chinese yuan-lin reflects the lifestyle of the literati; it does not address the environmental and social problems China is facing today. To respond to contemporary issues of environment, energy and population, we must seek for appropriate approaches. To see traditional Chinese yuan-lin from this point of view, then one will find: first, it is expensive; second, it is private instead of public; third, it lacks locality; fourth, it lacks environmental and ecological consciences. Yuan-lins are poetic but do not confer with contemporary values. We must have higher standards in measuring beauty and ugliness. We need a new aesthetics—this I call it the "down-to-earth aesthetics" or "low-carbon aesthetics".

WU: You think you have found the new vernacular, don't you?

YU: I think so, a new vernacular that fits contemporary China, addresses environmental problems and serves the Chinese people.

WU: Does this "new vernacular" equal to urban feng-shui?

YU: No completely, in fact feng-shui and yuan-lin are two different systems in China. The design of yuan-lin never uses feng-shui; geomancer was not involved in yuan-lin construction. Although feng-shui was applied in imperial constructions, it is derived from commoner's everyday practices in site planning of tombs and houses. It is fundamentally a vernacular matter, a practical "art of survival". The vernacular meant everyday, living and functional. While yuan-lin was an art of leisure belong to the literati class. These are two different cultures. I come to think that feng-shui is about the culture of human survival and adaptation to the environment, while yuan-lin is a pure form of aesthetical creation.

Feng-shui originated from pre-scientific era, but I believe landscape design should share the same ideal, that is the harmony and reciprocity between human, land and nature. Feng-shui is tied to specific vernacular environ that registers the imprint of its people's adaptation to the land. It is a functional and daily need created by user groups. In our designed landscapes, including urban developments, one must take into consideration China's environment and energy challenges, food supply, and global warming, etc. All these are the essential aspects in defining the "new vernacular". The old vernacular came into being through generations of Chinese people's life and work on this land. The new vernacular is a response to contemporary environmental challenge, a must in order to deal with the survival of a huge population facing a fragile ecology. The urban environmental of China today apparently cannot afford the lifestyle of ancient literati.

The new vernacular is the most efficient and economical way to follow the local climate patterns and to secure human living conditions. Because the new vernacular is the simplest, the most feasible, it is unique. The new vernacular must be from the land itself, so it uses the least amount of natural resources. Therefore, we must understand the concept of "new vernacular" against contemporary backgrounds.

There was also the issue of classification. The old vernacular was not considered high culture, not considered as cultivated. In the past, in order to pass from the vernacular to the sophisticated, one often chose to take the paths of civil exam, urbanization or westernization. As a result of such thinking, a great deal of non-vernacular things filled contemporary Chinese cities, such as Baroque landscapes. Westernization and modernization are not the same concept. Modernization does not only belong to the West, it can certainly be derived from the vernacular. It is totally possible to be both modern and Chinese. In fact, this is a post-modern perspective, stressing the local identity. There is criticism about my landscape design as Western. It precisely confused concepts of modern and Western. Modernity does not conflict with local identity; the modern can also be the vernacular. On the other hand, if one mixes the tradition with the local, then he simplifies both concepts. Opposing what is modern and what is Chinese is the biggest mistake in many people's mind, a huge perceptual obstacle. It is insufficient to depend only on traditional Chinese style and forms. Can modernist style produce Chinese design? of course.

Actually, what we are discussing here is the issue of being new and Chinese. To me, there is no need to search for Chineseness from the old China. As long as the design solves Chinese problem, addresses Chinese people's daily needs, uses currently available Chinese technology, suits China's specific environment, and is made by Chinese people, then why shouldn't it be considered as Chinese? Of course it is. This is the case of my project Red Ribbon and Qijiang Park, both have addressed these issues.

WU: The Red Ribbon has a modernist design taste. I know you worked at the office of Peter Walker and Sasaki. What is your opinion of Peter Walker's design?

YU: Peter Walker is a key figure of modernist landscape. He introduced into landscape design the simple style of modernist architecture and modern art. His minimalism demonstrates whole simple forms. All his elements are geometrical and such geometry is in accord with modernist architecture. This is his great achievement. He focuses on modernist forms, transforms minimalism in architecture and art into modernist gardens. But he does not interested in ecology, rarely does he uses wild grass or other formless things. The fences in his design need to be pruned every year, the rocks settings are refined—any change would have destroy the formal beauty.

Pater Walker's influence on me was the beauty of his forms, a kind of simple and clear beauty. In my work, you will discover that I use different materials, the contents are also very different from his. Ecology and the ecological service function of the landscape are in core of my design. However, my emphases on ecology are not simply a return to naturalism and environmentalism, but to unite ecology, productivity and aesthetics, as well as environmental ethics and art. Some thought the form of the rice paddies on the campus design of Shenyang Architectural University is similar to Walker's design for Kavinsky Hotel in Germany. In fact they are totally different. The campus design presents a landscape that is productive, sustainable and collects rain water. It values the vernacular landscape, its productive function and ecological service. In reality I might be more close to McHarg. But I am also different from McHarg, but including art. Environmentalism and modern art appears both in my design. I also try to have my own creation and character among international environmental designs. For example, the Qijiang Park differs from both the Gasworks Park by Richard Haag in Seattle, and the Duisburg Park by Peter Latz. Both projects produced in the West don't include much new designs, mostly preserving the existing, treating pollution, exhibiting old machinery… what was presented was mostly environmentalist perspective. Qijiang Park not only has ecological restoration and environmentalist ideal, but also stresses new designs and new experiences, such as the design of railroad and paths, the invented red box. Art is introduced at the same time of environment and ecology. The use of wild grass also gives a hint of the post-modern. The Yongning Park not only solves flood problem but also is artistic design. So my works benefit from minimalist design masters like Peter Walker for their artistic inclination, but their content are influenced by contemporary environmentalism and eco-ethics. There are minimalist forms, there are modern art, but all exist against the background of ecology. So, I prefer to call my design as Ecological Minimalism. Planting wild grasses and restoring wetlands are good, but those are merely ecological principles. Design has to be artistic, to have contemporary thinking. I emphasize the coexistence of ecology and culture. When ecology, culture and art are in one in design, one produces the true contemporary landscape. And this is where the new vernacular of China dwells.

Only when these two sides are combined, one may achieve unique characters of today's China. Only following ecology, then Chinese landscape will lose its cultural specificity. It would only manifest the natural process, just like what Peter Walker called *"invisible garden"* with no cultural meaning. We have to have both. A design without art, then there is no humanity and significance, and this profession would have no point of existing any more. In the book *Invisible Garden* Peter Walker in fact criticized pure ecology, promoting the importance of design and art in landscape creation. What I am doing is to add to his call a renewed attention to ecology. This is not any more Environmentalism in its old meaning, but the marriage of art and ecology.

Turenscape's recent project Tianjin Qiao Yuan demonstrates an ecological minimalism. Small bubble-like concaves were dug on an abandon ground to collect rain water. Plant groups would then develop naturally around these ponds; later we insert wood platforms for human observation and experiences. The ecological beauty of the bubbles is thus presented; it is as much art as ecology. Ponds in different depth would have different water levels and chemicals, so the plant groups would vary. These are all science. But the arrangement of the ponds and depth are totally artificial. This project is the opposite of the Red Ribbon. Qiao Yuan inserts ecology into art, while the Red Ribbon plugs art into environment. Both are very efficient.

The greatest challenge to China now is environment, and the key to solve environmental problem is to rebuild the tie between human, land and water. This is what should be focused on in urban landscape design. For instance, it is nice to reveal the circulation of water between rain and underground, making the natural process visible. Ecological infrastructure is to rebuild the relationship between human and land by reconstructing the rain water cycle in urban areas and reviving native plants, cultural tradition and daily activities…

In fact, Europeans and Americans are also reflecting upon such issues. Architecture turns green, showing a parallel with landscape design. The showy architecture and structure by Rem Koolhaas cannot be continued. Eventually, we need to turn towards ecological and down-to-earth architecture. It is the same in landscape design: formalism would have to given way to ecologically conscious approaches, which is what I define as the large-scale landscape security pattern of green cities and country.

WU: These enter planning domain. Can you explain your idea of "Negative Planning"? Is it still urban planning, but merely a different approach?

YU: Yes, negative planning is a new approach to urban planning, based on reflection and improvement of traditional urban planning methods. It is developed from my understanding to present issues in China and the limits of current Chinese urban planning administration and methodology. "Negative Planning" is not needed in Europe, since majority of the urban planning are done and the development of many cities are complete. "Negative Planning" addresses the reality of Chinese cities where the speed of urbanization and development supersedes urban planning. No matter how nice the plan was, there is little chance for it to take shape. So, why not first make a large-scale "non-buildable area" planning, an unchangeable preventive planning? Wetlands, rivers and cultural sites are easily ignored by developers, so I protect them first, by planning by-law. Thus buildings maybe built and demolished in decades, but ecological corridors and framework remain unchanged. This is "Negative Planning"—it eliminates possibilities of big mistakes in development and construction. Nothing can be more useful at this time of unprecedented urbanization in China.

WU: How can you guarantee that the local government would follow through?

YU: Through discourses. I learnt this from invited lectures to mayors. During these lectures, I stress the importance of "Negative Planning". It may be integrated into the master plan, but its outcome must be established as law prior to any physical planning. "Negative Planning" does not limit development, it only tells where cannot be occupied. This is an acceptable base line to anyone—the protection of 30% of area meant possibility of development in the rest of 70% area. "Negative Planning" is an invention according to the situation of China—rapid urbanism, fragile environment and unregulated development. But it also has universal significances. Traditional planning often aims at a perfect picture of the future while urban development is always unpredictable. So, protective planning is far more feasible than projective ones.

WU: Your office, Turenscape, promote a down-to-earth culture and the beauty of wild plants. What is your opinion about the fashionable agriculture tourism?

YU: Turenscape wants our design to promote a new aesthetics based on contemporary environmental- and ecological-ethics. Agriculture tourism gives experience to the art of survival. It is different from tourism to scenic landscapes, and represents a new tend in our society. It enables urbanites to return to the land, experience the productivity and beauty of the land. It is a new economy, representing a return to the land from urbanism. This also means that agricultural and productive landscapes will become an important domain of design in the future. Landscape design steps out of the unusual into the everyday life.

WU: Can you discuss how other designers can benefit from the experiences of Turenscape, but avoid felling into formal imitation?

YU: I think the lesson of Turenscape is manifested by its name "turen, earthman"—its love of the land and the passion of harmonizing man and earth, its unwaning emotion to the vernacular country. Many people thought there is a united style in our design. In fact, we do not intentionally pursue any style or character. If there is one, it is because we design for the characteristics of the site. No two sites are the same, just like two persons are different. Inspiration of landscape design lays in the land itself. If one follows this logic, there is no worry to fell into Formalism. Similar problems might have same or different solutions. If the problem is located in specific situation, then the best solution to the same problem would naturally vary. Of course, in order to achieve this, one has to be a devoted professional. So I believe the experience of Turenscape is not only its epistemology and methodology, but also, more importantly, its devotion and professionalism. We are not perfect in this aspect, there are often regrets, and we keep learning from our mistakes.

WU: After the Copenhagen conference, many have started to understand the complexity of our climate problem. Global warming cannot be solved merely by science and legislation. It is a concern a series of issues: energy and consumption, history and culture, economy and development, poverty and population, etc. Many insightful designers targeted our over construction and the method of construction, pointing out that human being must rethink what is design and design for what. If we don't change our fundamental understanding to construction, and construction would be destructive to the environment. No matter how green landscape architecture is, it is still artificial construction…What is your view?

YU: I totally agree. First, we built many unnecessary things, waste a great deal of natural resources. There are many empty cities in the world, same in China. Second, we built on places that should not be built, such as riverbank, wetland and ecological habitats. When destructive urban development destroy ecological infrastructure for the survival of species, we will soon see the consequences. Third, we built high maintenance architecture; an accomplished design today becomes a future burden. Similar problem is in landscape architecture.

Reducing CO_2, living low-carbon life and building low-carbon cities currently are the top challenges to human society. There is no magic solution. To defeat global warming is far more difficult than defeating Hitler, because it means to change human being's basic biological habit and thought. We are born with tendency of wasting, no concern of saving. In a environment abundant with resources, exploitation of nature was the driving force of human evolution and development. Then we realize all in a sudden, there is only one earth. So there is the dilemma between the unlimited consumption and the limited environmental sustainability. Some dream of solve the problem by being more energy efficient and reducing emission (such as in plane, car and housing design). The fact is that science cannot erudite global warming. As long as human being desires, any science or technology cannot be the solution but merely placeboes, just like high-speedy internet cannot reduce our office hour and wider roads only lead to more cars...

The key is human desire. Had human desire change from the infinite to the finite, there is hope in rescuing the earth. Obviously, we cannot ask most people to follow Buddhist practice, but we should have shared beliefs—to me, this is aesthetics. The pioneer Cai Yuanpei had advocated the idea of replacing religion with aesthetics. If all people have lofty aesthetics-such as realizing ornamental landscaping is indeed ugly, showy architecture is strange instead of beautiful, walking and cycling are more gentlemen than driving, neighborhoods planted with vegetable and corn are pretty, etc—then we would be very close to a low-carbon life, and this earth would be long lasting.

In summary, we must admit the number 1 enemy of a low-carbon world is human being's consumption and desire, science and technology cannot provide the perfect solution. And my prescription is to promote full-heatedly a new aesthetics and a religious love of the land.

WU: Your critical attitude is similar to the Chinese New Culture Movement at the beginning of the 20th-century. Mr. Cai Yuanpei studied abroad after first succeeded in traditional civil examination. I notice you are cultivated in classical Chinese. Can you explain your view of Chinese cultural tradition?

YU: This is a huge topic, hard to exhaust. Even though I stayed in the US for 5 years, my main education and thinking are bound to China; I have forgotten for a minute that I am the son of a Chinese peasant. About Chinese cultural tradition, my attitude is greatly influence by the New Culture Movement pioneers of the generation of Hu Shi and Lu Xun. I think there are two tiers of Chinese culture. There is the living low-culture, rich of variety and continuous up to now. This is the active energy essential to Chinese people's survival and development. In relation to the principle of landscape design, this vernacular culture includes how to tend the land, how to choose site, how to manage the water, how to dwell... For thousand years, this culture has provided for the survival of a big population within a limited land and resource and very different topographies. The other tier of Chinese culture belongs to the high class of literati and aristocracy, oriented towards leisure and unrelated to survival. Historically, we paid more attention to high culture while ignoring low culture. In order to face the contemporary environmental crisis, we have to promote the art of survival, the living vernacular culture, not the dead culture of the imperial and literati. At the present, many talk about cultural tradition and heritage, but often detach from the life of common people and confuse out-of-date practices as national essences.

WU: What is your understanding about aesthetic education?

YU: Aesthetic education is the base of all education. Mr. Cai Yuanpei wanted to replace religions by aesthetics in China, meaning aesthetics is as important as religious beliefs. I mention the beauty of wild grasses is aiming at the development of a new aesthetics. My article "A down-to-earth Culture and the Beauty of Wild Grasses" has been included into middle school textbooks. In fact, this is to use a new aesthetics to produce an impact that stimulates new ways of thinking. The students in a school in Huaibei were assigned to respond to my article by observing the land and its beauty. Their writings are very moving. They see the beauty of the wild grasses, the fish ponds, the rice paddies...such are the meaningfulness of the new aesthetics and the new vernacular. It is in fact related to the New Culture Movement, related to revolution in literature. Today's China needs such an enlightenment of new aesthetics. It is the continuation of the May Fourth Movement, the modernization of aesthetics, and the foundation of a people's progress. But it is not easy to make more people to accept it.

Landscape design is a social activity. We must educate policy-makers, developers, designers, builders and users. This discipline must be reframed, reevaluated and establish new standards. It is impossible to create good design when the basic value system is false. Therefore, one of the main aspects of our tasks in the recent years has been to redefine the discipline of landscape design, seeing it as an art of survival that derives from Da Yu's water management, and 5000-year experience of irrigation, field-formation and cultivation in the vernacular China. Defining landscape design as the art of survival can be thought as the continuation of McHarg's theory to a certain extent, but it is more the result of my rethinking of the history of environment and agriculture on this land of China, it is a response to both our ancestral experience and contemporary challenge. My thinking today benefits largely from my teaching and practicing in China after returning from abroad, but I believe it could also have global reach.

WU: I look forward to your producing more and more thoughtful students. You are the dean of the newly established School of Architecture and Landscape Design at Beijing University. Can you talk about your perspective of landscape design education? In the 20th-century, the field of landscape design has expanded greatly, and the problem of globalization has made more and more individual and countries to reconsider the cultural significance of landscape. Both modernist way of thinking and techno-science structure of the society have been challenged, while more and more attention is paid to cultural context and local specificity. Would you have any advice to Chinese and Asian students studying abroad?

YU: First of all, what I call Landscape Design differs from Feng jing yuan lin in traditional sense. Chinese yuanlin has an unchallengeable long and glorious history, but Landscape Design is still at its inception. We need to explore and improve, the system needs to be completed, and there is shortage of good teachers; but the future is very bright. This field must build upon the bases of natural science, social science and the humanities, as well as civil engineering. Land is the subject of own study, the harmony between man and earth is the aim, while the goal of our education is to produce designers who understands contemporary environmental ethics, masters modern science and technology, and is keen on negotiating the relationship of human and nature. A new aesthetics based on ethics is the foundation of this discipline. The art of survival is our primary point of departure. However, to develop such a discipline needs the effort of several generations. At the present, there is already the enthusiasm.

The passed decade has paved a hopeful way towards a bright future. In the coming decade, the focus would be to sort out the structural roots and network. The maturity of a discipline is just like that of a person, it needs a cycle. I think in ten year Chinese landscape design education would have reached a thriving and healthy stage.

As for my suggestion to Asian students abroad, first and foremost is to think about your homeland and the world. This is not mere rhetoric, but a kind of attitude and goal of study. Be clear that the aim of your study is to change the world while recarving your own path. For Chinese students, the most important would be to remember China today is facing tremendous challenge of survival, from environment to culture. So, cherish your opportunity, focus your energy efficiently, and do not waste time. Everything you learn will be useful one day; learning nothing then you can achieve nothing.

戴安娜·巴摩里
——城市景观与自然美

戴安娜·巴摩里（Diana Balmori）是一名西班牙裔的美国景观设计师，同时也是耶鲁大学的设计教授。1990年，她成立了致力于景观与城市设计实践的巴摩里联合事务所。事务所因为在创新的设计方案中融入了可持续系统，并使之在复杂的城市项目中得以实现而享誉国际。在她的设计理念中，景观首先是一门艺术。2006年，她的公司在事务所内部设立了一个叫做巴摩里实验室的机构，以探寻景观设计在思想和形式方面的发展。巴摩里实验室开展了广泛的研究，以搜索与特定的设计理念相联系的形式。这种形式的研究集中在寻找能够使景观和建筑、艺术以及工程等领域相互交叉融贯的途径。她最近的作品展示了城市空间中景观生态的新途径，他们通过采用先进的技术和大胆的形式，以及她付诸实践的生态原理而表现出色。她曾在亚洲主要的国家（如中国、韩国、日本和印度）做过项目，她的作品被翻译成多种语言出版，如英文、意大利文、中文和韩文等。巴摩里拥有城市研究的博士学位，广泛地讲授景观与城市空间之间的对话。她目前正处在美国美术委员会华盛顿分部的第二届任期中。同时她还撰写或合撰了许多书籍，包括《土地和自然开发准则：可持续土地开发指南》、《再设计美国草坪：寻求环境和谐》、《公园再定义：过去和将来的公园》。她的新书《景观宣言》由耶鲁大学出版社在2010年夏出版。最后，巴摩里博士一向认为景观是一门艺术，有助于同时从空间层面和社会层面治愈城市的病症。那本非比寻常的书《瞬性花园：无根的生活》是她与摄影师玛格丽特·莫顿（Margret Morton）的合作，意在努力呼吁关注纽约市无家可归人群的生活。她的团队最近在意大利和无家可归的儿童们一起，在"拯救儿童"的国际组织的帮助下，策划了一个艺术展览。

吴欣：首先，祝贺你的新书《景观宣言》的出版。我了解到，它包括了25条景观宣言，并将在2010年9月的ASLA华盛顿年会暨博览会上首发。我们的读者一定都期望阅读书中的每一条宣言。我们能不能从这本即将面世的书开始：你关注的主要问题是什么，写此书的主要目的是什么？

戴安娜·巴摩里（以下简称巴摩里）：写此书有很多想法，这也是为什么它提出了25条。历史上总有一些发生巨大变化的时刻；而我觉得了解我们所生活的这个时刻以及它提供的机会是非常重要的。它提供的机会是在人类与地球的其他部分之间建立不同以往的关系，无论是与植物、动物、空气、水还是土壤。直到20世纪早期，人类与地球上其他部分之间所划分的界限还十分清晰。人类在一边，自然在另一边，自然包括除去我们自己的一切事物。自此以后，由于多方面的哲学和科学原因，我们已经认识到，我们也是自然的一部分，并且我们的命运是和这一切紧密联系在一起的。然而，到20世纪末为止我们所做的一切，都基于这一陈旧的理解，而并不是出于什么邪恶的目的，或是企图伤害谁，而是因为我们按照过时的思维方式来理解人类与自然之间的关系。所以，现在我们的行动终于代表了新的理解，但行动的导则尚未确定，导则必须要从这一认识中继续深化而来。因此，我们需要通过行动来建立一种新的认识。我想谈一谈我们必须要跨越的3个不同的障碍。

第一条是人类和地球其他部分的关系；另一条则是城市和农村之间的界限。如果现在50%以上的人口居住在城市中，那么城市应该成为我们所要关注的中心问题。城市，因为是人类的创造物，曾被认为与自然毫无关联，因而从文化和物质的角度把它与自然完全隔离。因此，目前我们对人类作为大自然的一部分的新认识，立即影响到我们对城市的认识。城市要和其余部分一起融入到天地万物中去。这意味着，郊区、农村和荒野，所有这一切都需要以一种新的方式来看待。城市能够包含所有这些成分，并在改变它们的同时也为它们所改变。这种理解意味着我们修建城市的方式必须发生很大变化，我们不应该从根本上再将城市看作一系列的建筑和道路，而应该将其看作一系列的土地系统，相互联系，互动运作。

我们现在正在做的事情——我们正在做的绿色屋顶，我们正在做的绿墙，我们正在创建穿越城市的线性公园，以在城市中呈现自然的绿廊，呈现宇宙中有生命的部分。但是我们所做的只能算得上是皮毛而已。我们必须充分认识到城市不是由与自然相分离的碎片所组成的，它必须与其他系统，包括水、土壤、空气等作为一个整体共同工作。因此，我们可以给出一个例子来说明如何开始转变思维方式，设计和制造城市，例如，通过绿色屋顶这种非凡的方式。

不过，如果你真的希望看到屋顶绿化作为一种理念是如何引致更广泛的领域，我可以谈谈我们为韩国世宗特别市所做的一个国际竞赛获胜方案。它基本上将屋顶——在那些5层楼的屋顶之上改造出一个连续的"第六层"层面，将其转化为公共空间，成为城市公共生活的支持。所以，屋顶不仅仅是用做种植绿化的地方，它也可以成为充满活力的公共空间，成为连接城市的不同部分，对那里的居民来说，这也赋予了将城市浸入到自然的新的含义。这是对待城市的另一种截然不同的方式的开端。

第三个障碍是专业之间的界限，正如城市和农村被认为是完全不同的两个世界一样，不同专业也将自己封闭在自己的小圈子内。建筑负责在城市中建楼筑厦，景观则负责郊区，最多也仅是在城市中负责建筑的基底周边设计。而在新的视角下，这两个行业间的界限正变得更加复杂。事实上建筑现在很大程度上受到了景观的影响，而且现在已开始大量仿效景观的做法，使用等高线、使用摄影，并试图在楼宇中模仿生命过程，因为景观中一直都涉及不断的生命变化，而这是建筑中从来没有的。因此，我们正在跨越一系列的阶梯或边界，来修正这些专业之间的关系。这种跨越一旦完成，我们将身处一个完全不同的世界。这就是这本书所要讲的一切。它实际上所要讲的是这一变化的时刻，这一跨越的时刻。它是我对这一时刻所能给予的最准确的描述，是对在我们对自身新的认识上如何取得进一步发展的描绘，是对已成为我们生活焦点的城市的描绘。

最后，纵观今天可以影响人们生活形态的行业，我认为景观设计无疑会是我们从一个世界走向另一个世界这一转换中作出重大贡献的行业之一。

吴欣： 谢谢你如此详细的回答。那么景观设计的重点应该是我们现有的城市？

巴摩里：是的，这是现在唯一紧要的问题。50%以上的人口居住在城市中，而且这一比例正在飞速攀升。正在发生的事情是城市不能容纳这么多蜂拥而至的人，它现在给为数庞大的人口提供的是一种梦魇般的可怕生活，没有任何保障、没有水、没有电，没有任何方式能体面地让他们为自己建造一个一砖半瓦的容身之所。因此，不同的城市正在试图用不同的方式来处理这个问题，但事实摆在那里，一波接一波的新移民涌入城市，这些移民没有办法享受到城市的服务或是找到一片屋檐可以挡风遮雨。因此，我们真正地将注意力集中在已经成为21世纪生活中心的城市是至关重要的。

吴欣：让我们回到关于专业转型的问题上。当在描述你自己的大型城市景观工程时，你常常呼吁创建一种"更模糊的边界"。你的意思是否是要试图建立一个新的领域，一个融合了景观设计与城市规划的新领域？还是你在探寻令景观设计与城市规划各自拓宽其专业领域的方法？

巴摩里：我认为是后者。该想法来自于在处理城市问题中——特别是在总体规划和城市规划中，现在的景观干预手段和我们之前所采用的方法有着很大的不同，之前我们所采用的方法基本上只在土地表面工作，在如此大的尺度上，它往往只停留在了分析图阶段。景观要求在好几个层次上工作，从地上的大气层直到地下的地质层。并且它还要求在界定更加严格的尺度下进行设计，因此与分析图不同，景观对城中的每一块土地间的细微区别都提供了更加综合全面的理解。因此，景观不是把城市看作是一个个相似的平面，而是认识到每一片土地都是不同的，它下面的土壤是不同的，一些地方有水汇集而另一些地方没有，所以这开始显示出城市内部的差异，而不是所设想的同质性。因此如果景观通过一些手法将差异呈现出来，创造出一个与地形、气候、条件、朝向相适应的城市，景观的定位将要出色很多很多。

当然，我并不是说城市规划就会消失，而是说景观所采取的手法对今天的城市规划来说是必不可少的，而且需要处理好建筑的朝向，也可以使用地热来给大厦供暖或是利用某个方向流动的风来调节整座楼的空气流通,并降低能源消耗,来解决其自身的问题。那就意味着所有的一切都更加依赖于它的区位、朝向、部署和对所有这些层面的分析。

吴欣:你的手法是如何联系到最近很热点的"景观都市主义"?

巴摩里:它已成为一个非常实用的词,因为它主张把景观应用到城市中。但我非常担心它的另一面,这就是为什么我总试图避免使用这一术语。让我担心的就是它暗含的是都市主义,这将景观自身与最大尺度上的都市主义联系在了一起,在这一尺度上你是在画分析图,而不是在设计不同的城市地块。它更多是在同化,而不是在异化。在我的思维当中,这是一种属于总体规划时代的工作方式,如果景观陷入其中的话,会令我非常担忧。这是一种形式的都市主义,但我认为它已经死了,我们不应该进一步讨论它了。它意味着在上部鸟瞰城市,而不是分析城市内部各个部分的差异。因此,我希望景观都市主义这个词不再推着我们向更多的城市分析图的方向走,而是把我们的注意力转向对城市每片土地上最小尺度的差异性的具体设计。

吴欣:你所说的对于景观界来说非常有见地。我很期待你的新书,它意味着景观设计为规划带来的观念更新。你作为一名教授,在耶鲁大学的教学中,以及在你自己的实践中都很努力地去提高艺术在景观中的重要性。这似乎为你的设计引入了一个完全不同的层面。

巴摩里:是的,在这里我要澄清一下,因为当你在城市尺度工作时,人们认为在某种程度上你是在处理技术问题,本质上你是在解决问题,如生态问题,或是工程问题。可以肯定的是,城市有很多需要解决的问题。但我的确认为,在做这些城市发展工作时,你的动力,必须是出于审美的。你应该照顾到审美,从而给人们营造愉快的居住环境。这点绝对是必要的,也应是在做项目时首先考虑的。紧接着,你从技术、工程、生态等方面做出一定的贡献,但主要的动力必须是来自创造可以带来乐趣的东西。然而,不知何故,景观已经走出了艺术的大家庭，因此，举例来说，博物馆不认为他们应该举办一个景观展，因为景观不是一门艺术，他们认为它一方面或多或少是属于园艺界，又一方面或多或少是属于规划界的。因此，景观设计被完全误解了。但整个行业的过错在于一直没有展现出其在艺术方面的能力。它被艺术界驱逐，成为生态、成为园艺、成为工程、成为都市主义、成为其他很多东西，却忘记了它是一门艺术。但我们应该记得伟大的景观之所以留存下来，是因为它们的艺术性，因为技术好，因为工程好，并且结果是一个令人愉悦的美好的事物。这是出发点，其余的取决于我们。

吴欣：但是我们不得不承认，景观与在博物馆中通常陈列的艺术品是不同的，这可能是景观作为艺术的公共认知问题的核心。景观设计在技术设计和艺术形式之间的权衡面临着极大的困难。你能阐述一下你自身追求艺术形式的方式吗？

巴摩里：你说的完全正确，没有人真正知道如何应付景观作为一种艺术的难题。但是，这部分原因可能是因为景观艺术家并没有像展览艺术品那样展示他们的景观作品。他们也没有找到适当的表现系统。景观设计师所使用的表现系统相当简陋，无论是他们的图像还是他们的模型都很贫乏。建筑师在这方面更加在行。但是，建筑师和景观设计师都面临着越来越多的表现他们作品的问题。原因有三：第一，电脑的出现，改变了基本的表现形式；第二，计算机的表现方式影响了设计形式，模糊了建筑、雕塑和景观表现之间的界线；第三，表现形式自身的交流对象并不是一般大众，所以当前的表现形式不可能传递美学信息。当前的表现形式和普罗大众对它们的理解之间被划开了一个深渊。

所以，你的问题实际上需要我们思考两个问题：第一，新的表现形式的需要；第二，雕塑、建筑和景观之间的区别变得模糊。景观既不是雕塑，尽管它可以雕塑化；景观也不是地形，尽管它必须有地形。对我来说，艺术形式的独特性是与你如何使用空间、塑造空间和刻画空间相关的。你如何使用它、你如何使之开放并供人使用、你如何表现所有有助于每种艺术特性的东西。此外，如果你希望你的设计形式有助于产生异质性，吸引不同的人做出不同的解释，设计的要求就会更复杂。形式在我的作品中是空间，而非造型。

吴欣：这确实是一种艺术的挑战！设计中的形式应该永远不会仅成为一个符号，也不应该被预先包装以供视觉欣赏！你强调空间的使用特性意味着对于景观形式的认识不应该仅仅是关乎视觉的，而更应关乎生活体验。视觉图像在过去曾经是，在现在也依然是，并希望持之以久。请问你所倡导的艺术形式是否暂不考虑这一愿景而更关注当下呢？

巴摩里：没错。你不是为将来设计，而是为现在设计。你不能指望所有的使用者能够以同样的方式做出反应，相反，你是要建一个空间把不同的人集合到一起，来提升他们的活力，他们对生活的投入，激发他们发现自己的方式去接触自然的不同方面。因此，相比于制定一个计划、一种行为反应、一个共同的关注，或一项期待的目标，空间更应该吸引公众参与到现在的发现之旅中。为了达到这一效果，你必须创建一个愉快而美丽的空间，使得人们愿意待在那里。只有这样，他们才可能被调动起来，并以你从不曾想到的方式开始探索。人不会对世界上那些理所当然的事情感兴趣，反而只钟意于那些出人意料的事。因此，景观设计，为了激发疑惑和探索，不应该将其出发点定在确立要素之间的关系上——诸如土壤、水、河岸等宏观的概念差异之间的关系——而是应该创造模糊的"介于两者之间的空间"，这样人们就会疑惑他们到底是在水上还是在岸上。这样一个简单问题的思探，会引起每一位游客以他们自己的方式对于要素重新认识，可以促进他们用不同的方式对场所进行使用。它也可能使得人们接受水和河流的潮起潮落。这可能也有助于避免把涨水看成一种潜在的灾害，而要高筑河堤以保护自己，而是使人们可以在日常生活中与自然中万物的起伏共同生活。而这些新设计的场所要创造性开拓自然中那些我们认为是理所当然的方面，令人享受到无尽的愉悦和美丽。

吴欣："美"作为触发器？

巴摩里：是的，作为一个触发器。要探索一个点，能够刺激你对于自身文化中某些方面有关键性的重建，使你不得不被它所吸引，你需要能够留下来，庆幸能够融入其中。你要好好享受一下与场所的实用功能毫不相关的美的愉悦。这是为什么美是一切的支柱。美就是你的出发点。

吴欣：这种美学至上的思想是怎样引致你的作品中表现出诸如"生态基础设施"或"线性公园"这样突出的设计理念的？

巴摩里：它们是干预城市的工具。如果不把公园看作是一种表现形式、一种审美对象的话，你就可以感受到线性公园可以怎样穿梭于城市之中。它是一个用来衔接完全不同的街区的工具，鼓励市民相互融合。这样的走廊不仅能将人连接起来，也连接起了动物和植物。生物学告诉我们，廊道的连接性对于不同的动物和植物物种来说绝对是至关重要的。生物物种集群不应该被误认为是孤立的区域。甚至植物也有它们自己的扩散方式。因此，廊道是一个生物或生态王国，已被科学地证明是生物物种生存的重要工具，但是它也可以成为城市内的审美工具。你可以使用廊道创造最非凡的城市漫步道、连续的人行走廊。我们在城市中将这些本应该利用起来的空间完全让位给了汽车，但通过不同的途径，我们开始沿着河流的两岸，在废弃的铁路线、在活性廊道中重新发现、重新创造。这是一种新的基础设施，一种生态健康、意义重大的基础设施，同时它是一种新的审美机遇，一种创造别样公园的机遇。

吴欣："基础设施"是否建构了新的景观创意的出发点？

巴摩里：这是一个非常有趣的问题，请允许我解释一下我们对于人与城市中的自然关系的新认识是如何导致了对于旧的城市实践和语汇的新认识，并回过头来为景观创作打开了大门。"基础设施"是能够把一座城市中的事物联系起来的一个运作系统的名称，诸如水、气、电以及水管理（排水）等。这是任何发挥着某些自然方面力量的工程系统正常工作的保证，进而确保了城市的运转。它们中的每一个对于维持城市生活都起着重要作用。由于它们连接了人类的关注和自然的力量，它们可以成为我们手中的工具，以根据其自然特性区分城市中的各种部分。到现在为止，城市管理者一直认为电力、燃气、水系统等要集中，所有的城市系统都要集中。这些都是非常大的系统。但事实上，最近的技术表明我们也可以用较小的单位分配水电，这能使我们在城市中创造更大的异质性，以更好的方式满足当地居民的需求。因此，我认为我们正从19世纪工程基础设施的概念中脱离出来，工程师们据此声称它是唯一有效的方式，而我们正把它们打破成不同形式的系统，以顺应不同的生态情况。这种新的"基础设施"努力使城市的不同部分能按其自有的方式工作，而不是全城以同样的方式运转。

线性公园是城市基础设施的一部分，在于它创造了动物和植物的生命廊道，因而它实现了一个重要的生态功能，这一功能已被城市生活与自然的分离而破坏掉。同时，它也给予了我们一个机会，来进行不同类型公园的审美创造。

因此，对基础设施的严格审查搭建了一个平台，让我们用一种完全不同的方式来对待旧的事物。因此，我们可以同时改变基础设施和公园的概念，因为我们知晓了创造城市基础设施的新技术，并可以通过适应新的城市中本土自然的特殊性的方式来创造城市基础设施。我们看到一件新的基础设施时可以说："哦，一个公园！"这意味着它们会和我们以前所知道的任何公园或是基础设施截然不同。我们甚至改变了公园这个概念本身。在美国，以及世界上其他许多地方，一提到公园就会让人想起深入我们脑海中的奥姆斯特德式的图像[1]，我们也认为这是理所当然的。但是就在窗外的这个高线公园，就是一类新型公园的优秀范例[2]。而如果你走到另一个街区，你会看到哈得逊河公园，那又是另一类线性公园[3]。因此，基础设施为我们打开了大门，让我们接受旧的观念并改变它们。

---

[1] 戴安娜·巴摩里所指的是由弗雷德里克·劳·奥姆斯特德（Frederick Law Olmsted），他的公司，及后来他的儿子们所创立的公园系统，这是美国风景如画的城市公园的最终参照物。纽约市的中央公园（1857—1873年）和布鲁克林展望公园（1865—1873年）为城市公园在纽约市和美国的发展和流行作出了巨大贡献。

[2] "高线"公园是一个新近的更新项目，就在戴安娜·巴摩里的办公室外，这个项目位于一条穿越曼哈顿南部的废弃高架铁路上。由詹姆斯·科勒（James Corner，景观设计）和迪勒·斯高夫迪奥＋连东（Diller Scolfidio+Rento，建筑设计）在2005—2009年期间，改造成了植被覆盖的购物中心。设计追随雅克·韦尔热利（Jacques Vergely，景观设计）和菲利普·马蒂厄（Philippe Matthieu，建筑设计）1987—2000年间在巴黎设计的"林荫散步道（Promenade Plantée）"这个著名的范例。

[3] 哈得逊河公园是一个新近的项目，它由沿着从第59街到南部的炮台公园的绿道中的一系列的公园和项目组成。它始于1998年，许多景观设计师和建筑师参与了这一项目，现在仍在进行中。

吴欣：这种城市基础设施设计中的概念模式转换很有意思。但它看起来是要预言一个新的普遍合理性，据我所知你和亚洲国家，如日本、中国和韩国的公司已经合作很长时间了，你并没有忽视文化差异。你能否谈谈东亚的设计师们都是如何借鉴你的思路，并兼顾自己的文化的？

巴摩里：这是最难回答的问题之一，你说到了非常重要的一点。但我仍然认为，我们在世界各地的城市中遇到的主要问题是相同的：令人难以置信的快速人口增长；旧的城市基础设施完全不能支撑人口增长；在不断恶化的气候条件下，对公共空间和健康的城市环境的日益增长的需求。这些问题，对于所有人来说都是新的——不仅仅是亚洲或美洲——并且需要新的解决方案。我们正处在这样一个整个地球生态的转折点，并面临着令人难以置信的快速的气候恶化，因此，答案必须是全球性的。因此，在这个层面上，我认为答案对于所有人来说都是新的，现在，当你着手设计后，首先要做的则是如何联系当地的文化，如何适应当地的文化。我认为，唯一的好办法是，一定要对当地的材料非常敏感，而且，最重要的是，能够与当地人合作。城市体系的设计需要团队合作，这些团队要形成一种跨文化的沟通方式。无论我在现场做多少研究，也绝对不可能完全地了解另一个地方的文化，我确实需要有人来就当地的文化对我进行指导。我很幸运，当我在日本工作时，一位同事就日本文化方面对我进行了教授、帮助。而现在我在印度也试图开展一些类似的文化指导工作。我认为对这个问题而言，跨文化的团队是必不可少的，除此之外，别无他法。

然而，我们面临的问题在各地都很相似，但我们需要对所有人来说都很新颖的解决方案。因此，举例来说，线性公园就是一个新的孕育公共空间的方式，这在世界上任何城市中都是行得通的，但它却可以采取完全不同的表达方式。它可以使用不同的植物，它可以加入不同的东西，服务于不同的文化目的。为人类、植物与动物创建廊道的概念是行之有效的，是全新的，因此，它取代了奥姆斯特德式的风景，但这并不意味着你不能像给项链加上宝石一样在这些线性公园上加上奥姆斯特德式的公园。

吴欣：你曾为韩国首尔附近设计过一个重要的新城，你能说说这些想法带给你和你的韩国同事怎样的新的城市和景观形式吗？

巴摩里：首先，我们赢得了一个新城设计的国际竞赛。当时的韩国总统决定将 8 个部委迁出首尔搬到那里，因为首尔像其他许多大城市一样，无法应对大量移民的涌入。于是，他们决定在离首尔相当近的地方建一个新城，高速列车 1 个半或 2 小时就能到，并将部分的政府机构转移过去，作为这座新城的经济基础。就像韩国常见的那样，位于 6 个不同村庄的塔楼建设全面展开。同时也进行了一些基础设施的规划建设，建立了一个包括高速公路和相应设施的超级网格。所以，我们以此作为出发点，在弧形围绕城址的土地和河流中引入了其他的连接系统，以打破加置其上的概念化的同质性。这不是乌托邦，而是用此地的兴建来说明如何回应不同地域各异的人类关注和自然条件。重要的是要表明此意。我们与一位韩国建筑师共同为第一幢建筑准备施工文件。为项目建设范围内的第二幢建筑征集方案的国际竞赛也旋即展开。这次被另一位韩国建筑师赢得，他目前正在准备施工文件。但这个新城设计目前被暂停了，因为下一届总统不希望政府部门搬离首尔。他建议，将这座新城市改建为学术新城，兴建大学园区。但由于它已被合法指定为行政城市，从一种性质到另一种性质的更改需要一定的法律程序。而那些程序尚未最终确定。

在为新城做总体规划前，我们把原原本本的地形和已布好的基础设施作为我们的出发点。因为在这种情况下，城市围绕一系列政府机构而展开，我们希望这些行政楼是平易近人的。因此，我们将层高限制在了 6 层，但与其他城市项目相反，我们希望在地面层与第六层形成一个连续的公共领域。这意味着需要特别把行政楼的第六层以及连接它的坡道做成一个连续的公园，在多个地方设立连接到 8 个不同的部委的通道。当然还有其他一些散布其中的文化机构和其他一些元素共同形成了融入场地的城市。此外，我们特别尊重这个场地被群山环抱的感觉，希望人们从城市中各个地方都能看到这些山峦。我们也坚持不把小山丘移出城，而要使它们成为城市的一部分。这违背了韩国的通常做法——韩国到处都是高山，城市建设者们通常铲除每一片起伏的地形，以便在平整的地面上建立一座新城——因此，我们保留了场址内的两座主山，使它们成为环境的一部分，成为真正的景观的一部分。然后，我们保留了场地里流动的水、沼泽和现有的稻田，确保稻田能够在城市中有一席之地，它们可作为城市水净化系统的一部分。因此，我们尽量地去兼顾它们的本质——它们的农业价值，同时融入它们能给城市带来的新的服务价值——它们的生态价值。所有这一切的目的就是使整个城市的生活变成愉悦的审美体验的源泉。因此，这个项目的目标就是要让村民在离家很短的距离内找到所需的任何服务，能够享受到宜人的城市公共空间，体验公共活动的亲近而不会有城市的疏离感，同时也能在自己居住的高楼眺望远山美景，或是俯瞰这座建筑在公园之下的葱郁而繁荣的城市。这是一座新城，一种新的景观。而我认为它为未来的城市提供了一些指导建议。

吴欣：我发现你对探索性研究的关注在景观设计师中独树一帜。请你介绍一下和你设计事务所融为一体的"巴摩里实验室"——你能否说一下它是如何参与到屋顶绿化的景观研究中去的？

巴摩里：绿色屋顶从一开始就是一个研究项目。我们并没有具体的客户，但我们认为应该坚持要做绿色屋顶，于是就去考察纽约市的那些特殊地段，然后问自己几个问题：在那个地区能有多少屋顶花园？又会带来什么样的效果？绿色屋顶的集聚效应会是什么样的？

因此，我们选取了长岛市一个地区，就在曼哈顿对面，皇后大桥（Queensboro Bridge）另一头的东河岸边。我们选取了这一地区，是因为这附近的行业使用了巨大的"饼状"建筑。这些3层的小楼有着大面积的平屋顶。此外，这一地区还有一些塔楼和小一些的住宅楼。我们就把所有的屋顶都绿化了。作为一种对整体现象的结果评价的方法，我们把所有的表面积累加起来，它的面积竟然有奥姆斯特德在布鲁克林所建的那个巨大的展望公园（Prospect Park）的一半那么大。我认为这是十分令人意外的，这意味着如果你这样做的话，你不用买地或是别的什么就可以创建一个公园。即使你只是做最简单的绿化，比如说"粗放型的屋顶绿化"，只需在7.6cm的土壤上撒上种子，你就将获得相当于一半的展望公园地表自然植被的效益。我发现，这很能说明问题。这表明，我们没有用正确的方式来思考绿色屋顶的问题,因为我们一直在各行其是。但是,一旦你把它们聚集在了一起,你就将知道它们将可以洁净空气,也可以影响城市空气的热动力。甚至,事实上,你有机会为每座建筑物的居民们创造空间,使之最终成为公共空间的一部分。那些巨大的产业建筑物很多是巨大无比的折扣商店。只要有微小的刺激,他们就可将他们的屋顶变成公园一样的公共空间,他们可以在其中出售户外商品或其他任何东西。这一切仅仅需要采取新的空间设计和利用途径。这只是我们事务所所做的研究中的一小部分。后来,我们联系了长岛市,觉得他们可能会对这个绿色屋顶的研究感兴趣。结果他们同意了,还聘请了一个教育机构来专门分析对绿色屋顶感兴趣的房地产开发商。他们的一些客户来了,其中一个是电视工作室,叫银杯工作室(Slivercup)。他们说这不错,想为自己的工作室做一些屋顶绿化,所以我们就做了;然后有一个现代家具制造商要求我们也为他做一个。我们甚至还获得了一些津贴来研究这些屋顶花园一年中的生态影响是什么,把我们所做的成果变成数字,实实在在地展示出来。因此,我们认识到我们所说的所有这些实际上远超过了对银杯工作室屋顶的试验。目前,我们正期待着将在2010年年底获取的有关更小型的屋顶的结果。虽然我们还不知道所有的结果,它真的是一个很好的研究和实验。

吴欣:这些研究会影响纽约市的建筑法规吗?

巴摩里:现在唯一发生的事情,一件非常好的事情,就是现在大家开始接受绿色屋顶了。所以,现在你可以因你的绿色屋顶而获得减税。这样它就鼓励了人们去建造屋顶花园。而这反过来又能够产生一种绿色屋顶的小型相关产业。当我们开始参与绿色屋顶这件事的时候,仅有一个安装人员和一个种植人员,而到目前已有7种不同的绿色屋顶建造系统。而那位研究屋顶花园植物的伙伴,他的业务已遍布全国,他在全国范围内销售屋顶花园的植物,甚至包括了南部。因此,绿色屋顶甚至创造了商机,更可以带来巨大的产业转型。

吴欣:换句话说,越来越多的人正在参与到其中。

巴摩里：是的，越来越多的人参与到其中。我敢说在纽约，大家对此非常感兴趣，可能已经覆盖了纽约10%的屋顶。我认为在几年后，20%的屋顶都可能成为绿色屋顶。没有人真正地计算过这个数字，我只是从打电话来征询建议的数字中推断的。现在任何拥有私人住宅的人，都可以自己建造绿色屋顶，因为他们可以咨询那些已经建过的人，从中得到了参考。它已经是众所周知的了。但是没有人对屋顶的设计多做思考。如果你只是种了一层非常薄的景天，这其实不需要解决什么设计问题——我的意思是指它是一个很小的设计。

吴欣：《再设计美国的草坪》是你多年前的研究出版物，但我认为现在它仍然对中国的读者具有特殊的意义。你能谈谈吗？

巴摩里：这感觉像很老的著作了，自那时以来美国出版了很多关于草坪的书，但我们的是第一本。我和两位来自耶鲁大学森林和环境研究学院的同事共同完成了这一研究。它是一项常常让我回顾的工作，因为人们重新发现它，希望获得更多的信息。我觉得它的核心是：如果当地的气候并不适合草坪，你就不应该建它。整个美国西部都不应该使用草坪，这是显而易见的。那里没有水，而草坪的用水量又是如此之大！另一点是，我们所开发的草坪的生态是非常脆弱的，它们的生物多样性并没有得到提升，相反，基本上已无生物多样性可言，因为他们只种了一种或两种草，并且他们还使用很多化学药物使草坪看起来很完美。这些化学药物杀死了草里的各种帮助保持土壤的小动物。此外，草坪需要人工施肥，大量的氮肥被喷洒到了草坪上和地面上。部分的氮向下渗透进入到地下水中，并最终渗透到河流中，造成藻类疯长，于是杀死了水中的鱼。因此，它有严重的后果，实际上美国的研究表明，草坪中使用的化学药物对于河流水系的危害比所有农业中使用的化学药物还要大很多。所以它对于整个环境是非常有害的，它浪费了水，还危害人们的健康，因为用在草坪上的喷剂是危险的，应完全避免其与皮肤的接触，但实际情况并非总是如此。因此，我们要处理的是一个为美观而建的景观，但这种景观并不是四海皆宜，因为它源于英国，那里频繁的小雨滋润的气候可以维持它的效果。首先，在没有雨的情况下，你不应该使用公共供水系统来维持它。

其次，在整个 18 世纪，直到 19 世纪末，大型的化学品和设备公司就开始对其进行掌控，草坪有为数庞大的植物品种可供选择。而在这些植物中，有的像三叶草，在盛夏由于高温死去时，本身就会产生氮，这样你就不必人工去添加氮肥。此外，这种形式的氮是能为植物根系所吸收的，并不会渗入地下水中。因此，如果草坪建在气候适宜的地方，它也会更慷慨地回馈其他环境。我们自 19 世纪末以来开发这种产业化草坪，我们应该了解它对于公共环境的害处是多么大。我也理解能让人在泥土面上与孩子玩耍或是玩儿童游戏的草坪的吸引力，因此如果我们确实要建草坪，我们也应该控制它的面积。我们应该在气候允许的地方建草坪，而我们植草时，应该遵循 19 世纪前的模式，采用混合的植物，包含像三叶草这样的能使其生态可持续的草种。因此，从某个层面上来说，我们的研究要求对当代工业草坪进行改造，但另一方面也意味着，在因雨水不足而会造成草坪不可持续的地方，你不应该去造草坪，因为它是不好的景观，它是有害的景观。因此，在景观中非常重要的一件事就是，了解每一片土地适宜种什么。生活在一个自给自足、变化纷呈的可持续景观中，将会更有趣，更美好，同样它也不再使整个世界陷入处处都一样的窘境。

吴欣：这是又一个关于美与审美的问题。

巴摩里：是的，我想补充一点，18 世纪在英格兰建草坪的人的确是将它作为了审美的对象，这就是为什么它吸引了大家。因此，我们今天的任务是使新的草坪成为一个转变，成为一个审美对象，而不是追求产业化草坪的美。因此，我们也需要重新确定设计策略，让它会像 18 世纪的草坪一样具有吸引力。然而是与那些产业化草坪一起建的艺术作品、大公园和花园，会使得草坪如此不可或缺，因而我们要同时建设这些相应的配套部分。

吴欣：谢谢你关于美学思想和生态思维关系的精彩论述，以及与中国读者分享你的设计思想是如何在自然界大的环境问题和具体场地的人与非人类自然关系之间穿梭往复的心得。　　（许婵　译，涂先明、吴欣　校）

Diana BALMORI
—between urban landscape and natural beauty

Diana BALMORI is a Spanish-born American landscape architect, and a professor of design at Yale University. In 1990 she established the landscape and urban design office, Balmori Associates, which is acknowledged internationally for its success in realizing complex urban projects that integrate sustainable systems within innovative design solutions. In her design philosophy, landscape is first and foremost an art. In 2006, the firm created BALMORILABS within the office in order to pursue the development of intellectual and formal aspects of landscape design. In BALMORILABS, extensive researches are undertaken in search of form related to particular design ideas. Such formal researches look intensively at ways in which landscape can intersect with architecture, art and engineering. Her recent works demonstrate a new approach to landscape ecology within urban spaces, excelling by their use of advanced technologies and daring forms as well as by the new ecological principles she puts into practice. She has worked in major Asian countries (China, Korea, Japan and India) and her work has been published in English, Italian, Chinese, Japanese and Korean. Dr. Balmori, who holds a doctoral degree in urban studies, has lectured broadly on the dialogue between landscape architecture and urban spaces. She is currently serving her second term on the US Commission of Fine Arts in Washington DC. She has also authored and co-authored many books, including *The Land and Natural Development (LAND) Code: Guidelines for Sustainable Land Development*, *Redesigning the American Lawn: A Search for Environmental Harmony*, *Park Redefinitions: The Once and Future Park*. Her newest book *A Landscape Manifesto* is being published by Yale University Press this summer. Finally, Dr. Balmori has always believed that landscape is an art that can help to heal urban problems, spatially and socially. A remarkable book *Transitory Gardens*, *Uprooted Lives* is the result of her cooperation with photographer (Margret Morton) in an effort to call attention to the life of the homeless community in New York City. In a recent occasion in Italy, her team prepared an art exhibition together with homeless children with the help of the international organization "Save the Children".

Xin WU(WU hereafter): Firstly, congratulations for your new book, *A Landscape Manifesto*. I understand it includes 25 design statements and will be launched this September at the ASLA 2010 Annual Meeting and Expo in Washington DC. I am sure our readers are looking forward to reading each of the manifestos. Could we start from this forth-coming book: what are the main issues you want to address, and what was your main purpose when writing it?

Diana BALMORI (BALMORI hereafter): This book pursues many aims, and that is why it raises 25 points. There are always moments in history when things change dramatically; and I find it imperative to understand the moment that we are living now, and the opportunities that it presents. It presents the opportunities of setting up a different relationship between humans and the rest of the planet, be it plants, animals, the air, the water or the earth. And a sharp edge had been drawn between humans and the rest of planet earth until the early 20th century. Humans were on one side, and nature on the other, and nature included everything but ourselves. Since then for several philosophical and scientific reasons it has become understood that we are parts of all of that, and that our destiny is tied to all of that. Yet, everything we have done to the end of the 20th century, had been based on the older understanding, not out of an evil purpose, or an intent to harm things, but because we interpreted the relationships between humans and nature according to an outdated system of thinking. So, now our actions have to represent that new understanding, while the directions of action are not yet established, they have to proceed from that understanding. So we need to cross over to a new way of thinking in action. I would like to speak of three different edges we have to overcome.

One is the relationship between human beings and the rest of the planet; another one is made of the separations between the places that we live in. If, by now 50% of the human population is living in cities, we are in a moment when the city becomes central to our preoccupations. Cities, like human beings and because they were a human creation which were considered to have nothing to do with nature, and therefore it was kept separated from nature from a cultural and physical point of view. So, at present, our new understanding of humans as part of nature affects immediately our understanding of the city. Therefore the city has to be woven in to the rest of the universe. That means that suburb and countryside and all that have to be thought about in new terms. The city contains these pieces, and it has been transformed by them as much as it transforms the rest. That understanding means that our way of making cities has to change dramatically, that we should no longer consider the city essentially as a set of buildings and roads, but rather as a set of earth-systems, that have to work in connection with one another.

The things we are doing now — we are doing green roofs, we are doing green-walls, we are creating linear parks that go through cities and therefore present green corridors of nature, of the living parts of the universe within the city. But these are tiny gestures still. We have to fully understand that a city is not made of places apart from nature, but that it has to work as a whole with other systems, including water, soil, air. So, one can give an example of how this begins to transform ways of thinking, designing and making the city, by turning to the very modus move towards green roofs.

But if you really care to see how the green roofs as an idea leads to a broader realm, we may turn to our proposal for a new city in Korea. Basically it turns the roofs, over all five story buildings, into public spaces as a continuous sixth floor level that becomes a support of city public life. So roofs are not simply places for planting greenery, but active public spaces which connect different parts of the city, and give a new sense of the immersion of the city into nature to its inhabitants. This is the beginning of a very different new way of approaching the city.

The third edge is the line between the professions, just as the cities and the countryside are considered totally a part of different realm; different professions had enclosed themselves in their own little edges. Architecture dealt with buildings in the cities, then landscape architecture dealt with the suburbs, and at the most in the city with the base of buildings. Within the new perspective the edge between these two professions is becoming much more complex. Architecture in fact is very much influenced now by landscape and it is beginning to ape lots of landscape practices, using contours, using photography, and trying to imitate living processes with buildings, since landscape always had to do with constant living changes, while architecture never had. So we are crossing a series of tiers or edges, revising the relationships between these professions. Once done this crossing, we will be in a totally different world. This is what this book is all about. It is really about this moment of change, this moment of passage. It is as accurate a description as I can give of that moment, of what we need in order to go forward on the basis of the new understanding of ourselves, within the rest of the world, of the city which has become the main focus of our living.

Finally, looking at the professions today that can affect those forms of living, I think that landscape architecture is one of the professions which can make a very significant contribution in the switch from one to the other world.

WU: Thank you for this detailed answer. So, the focus of landscape architecture ought to be on our existing cities?

BALMORI: Yes, that is the only thing that matters now. 50% of the population is living in cities, and it is moving very rapidly towards a higher percentage. What is happening is the city could not contain so many people who are flocking to it, and it is giving an enormous population a horrendous life without any support, without any water, without any electricity, without any ways of constructing for themselves a decent way of having a roof over their head. So different cities are trying to cope with it in very different ways, but the fact is there, they all have vast rings of people who have recently immigrated to them and have no access to city services or a roof over their head. So it is critical that we really focus our energies on what has become the center of life in the 21st century.

WU: Let us return to the transformation of the professions. When describing your own projects for large cityscapes you often call for the creation of a "thicker edge". Do you mean that your approach reaches for the establishment of a new field that would coexist with landscape architecture and urban planning? Or do you seek a way of broadening the domains of each profession?

BALMORI: I think the latter. Ideas coming from landscape today about intervention in an urban situation differ from the tools we have had before particularly in master planning and urban planning, which basically work only on the surface of the earth and at such large scale that it remains at the level of diagrams. Landscape demands working on several layers, from the atmosphere above to the geology below the level of the earth. It also invites working at more tightly defined scales of design, so rather than diagrams it produces an integrative understanding of the differences of every piece of land in the city. So it does not take the city as if it were a flat plane that is all identical, but it takes into account that each piece is very different, the soil below is different, the water is present in some places while in others it is not, so it begins to show the inner differentiation rather than the supposed homogeneity of the city. And therefore it has the tools to bring differentiation out and to create a city that fits its terrain, its climate, its conditions, its orientation much, much better.

So, by this I do not mean that urban planning is going to disappear, but rather that the tools of landscape are essential to urban planning today, and that the issue of buildings also now has to deal with orientation and also perhaps use the energy below the ground to heat the buildings or use the wind that circulate in certain directions to circulate air through the building, and reduce energy cost. It is that everything is much more dependent on its location, its orientation, its placement and the analysis of all of these layers.

WU: How does your approach relate to "landscape urbanism", a term that has been much talked about lately?

BALMORI: It has become a very useful word because it advocates the application of landscape to cities. It has a side that worries me enormously, which is why I try to stay away from using the term. And that is that it implies urbanism and it ties itself a lot to an urbanism at the largest scale by which you are doing diagrams rather than designing differentiated pieces of city. It homogenizes more than it differentiates. To my mind a form of work that belongs to the era of master plans and I would worry very much if landscape would get caught in that. It is a form of urbanism, but I think it is dead, and we should not go any further with it. It implies a view from above of the city rather than analysis of the differentiation of pieces in the city. So I am hoping that the term landscape urbanism will not push us in the direction of producing more city diagrams, but rather that we turn our attention toward specific designs at the smallest scale of differentiation within every piece of the city.

WU: I think that is very insightful for the field of landscape architecture and I really look forward to your book. It promises a renewal of planning methods, yet in your teaching as a professor at Yale, and in your own practice you try very hard to promote the importance of art in landscape architecture. This seems to introduce an altogether different dimension to your approach.

BALMORI: Yes, and I want to make a clarification because when you work at an urban scale, people consider that you are in some way dealing with technical issues, and that essentially your are solving problems, such as ecological problems, or engineering problems. It is certain that cities have lots of problems that need to be solved. But I really think that the impetus for what you do, when working at city development, has to be an aesthetic one. You should care for aesthetics in order to make something pleasurable for people to live in. This is absolutely essential, and it comes first. Then, later, you call upon contributions from technology, engineering, ecology, but the primary impulse must come from creating something that can give pleasure. Yet, landscape has somehow grown out of the family of the arts, so that museums for instance do not consider that they should deal with an exhibition on landscape, because it is not an art, they consider it more or less to belong to the realm of horticulture on one side, and planning on the other. So landscape design remains completely misunderstood. But the whole profession has been at fault in not presenting itself in its artistic capabilities. It has been drawn out to become ecology, to become horticulture, to become engineering, to become urbanism, to become a lot of other things, and forget it is an art. We should nevertheless remember that great landscape creations have survived because they were artistic, because the technology was good, because the engineering was good, and the outcome was a very pleasurable beautiful thing. It is where we start from. The rest is up to us.

WU: One has to admit that landscape is different from standard art objects presented in museums, and that may be at the core of problems for public recognition of landscape as art. Landscape architecture has great difficulties trying to negotiate between technical design and art form. Can you explain your approach to art forms?

BALMORI: Well, you are absolutely right. Nobody quite knows how to grapple with landscape as an art. But that is partly also because landscape artists have not been able to present their work as an art. They have failed to develop proper systems of representation. The systems of representation used by landscape designers are quite poor, both in their images and their models. Architects have been much better at this. However, both architects and landscape architects are facing more and more problems of representation of their works. There are three reasons for this: number one, the introduction of computers has changed the basic forms of representation; number two, the forms of computer representation affect design forms blurring the lines between architecture, sculpture and landscape representation; number three, the forms of representation themselves are not communicating to the general public, so it is no longer possible to transmit messages about aesthetics in the current forms of representation. It results into a big abyss between the current forms of representation and their understanding by the population as a whole.

So your question really demands that we address two things, first, the need for new forms of representation; second, the relative forms of sculpture, architecture and landscape. Those distinctions between these arts become blurry and that they shift into one another does not worry me. But landscape is neither sculpture, although it can be sculptural, nor topography even though it has to be topographical. For me, specificity of art form is related to how you use space, shape space and portray space. How you use it, how you open it to use, how you represent all contribute to the specificity of each art. Moreover, it is complicated because I want to give a form that does not tell people what to do, that does not manage people, but rather invites different interpretations by different people. I want forms to contribute to the production of heterogeneity. So, art forms in my work seek to be special, and seek not to be an object, they do neither seek to be topography, nor sculpture. Yet landscape form maybe sculptural in order to achieve the end of being space and being heterogeneous.

WU: This really is an artistic challenge! Forms in your designs should never become a sign, should never be pre-packaged for visual appreciation! Your emphasis on using as a feature of space means that the perception of landscape forms should not be merely related to visuality but rather living experience. Visual images were and still are expected to last into the long term. Does the art form you propose shy away from this ambition and privilege the present?

BALMORI: Exactly. You do not design for the future, you design for the present. You do not expect all users to respond in the same way, to the contrary you are creating a space that brings different people together, that enhances their energy, their engagement with life, and energizes them to discover their own ways of relating to different parts of nature. So, rather than proposing an agenda, a behavioral response, a shared set of concerns or a goal to be achieved in the future, it rather invites the public to engage in exploration in the present. In order to achieve that result, you have to create a space that is pleasurable and beautiful, so people want to be there. Only then they can be mobilized and start exploring in ways that you had never thought of before. Human minds do not wonder about aspects of the world that are taken for granted, but only about those that are unexpected. So, landscape design, in order to stimulate wonder and exploration, should not take at its departure point such relationships as the enormous conceptual distinction between earth, water, and river banks. It should rather create ambiguous "in-between spaces", where people do not know whether they are on water or on the land. The exploration of such a simple question leading to some renewed understanding of elements by each visitor in her own way, can spur you to use places in different ways. It may also lead to an acceptance that water and river come up and down. It may also help avoid seeing any rise of water as a potential disaster requiring the river to be dammed in order to protect yourself, but rather help seeing how you may live with the fluctuations of everything in nature on a daily basis. And these newly designed places have to invite this creative exploration of aspects of nature we usually take for granted in a way that becomes enormously pleasurable and beautiful.

WU: Beauty as a trigger?

BALMORI: Yes, as a trigger. To come and explore a place that stimulates critical reconstruction of some aspects of your culture, you have to feel attracted to it, to be able to stay you have to appreciate being in it. You have to enjoy some aesthetic pleasure, some pleasure unrelated to any practical function of the place. That is why the aesthetics are the mainstay of anything. And that is where you start.

WU: How does this primacy given to aesthetics lead to such design ideas as "eco-infrastructures" or "linear parks" which are prominent in your work?

BALMORI: These are tools with which to intervene in the city. Instead of considering that a park is a destination, an aesthetic destination, you can see how a linear park can weave its way through the city. This is a tool to connect very different neighborhoods, to encourage people mixing with one another. Such a corridor does not only connect people but animals and plants. Biology teaches that the connectivity of corridors is absolutely crucial for different animal species and for plants too. Clusters of living species should not be mistaken for isolated islands. Even plants have their own way of travelling. So the corridor is a biological or ecological realm that has been scientifically proven to be an important tool for the survival of living species, but it can also be turned into an aesthetic tool within the city. You can use corridors to create the most extraordinary walks through cities, continuous pedestrian corridors. This is something we have totally lost to the automobile in the city, but in different way we begin to rediscover and re-create by the sides of rivers, on abandoned railroad lines, in activity corridors. It is a new kind of infrastructure, an ecologically sound and meaningful infrastructure, and the same time it is an opportunity for a new aesthetic, for creating a different form of park.

WU: Does "infrastructure" constitutes the point of departure for new landscape inventions?

BALMORI: This is a very interesting question, that allows me to explain how our new understanding of relationships between humans and nature in the city leads to a new understanding of old urban practices and vocabulary, and in turn opens the door to landscape creation. "Infrastructure" is a name for the working systems that put together something like a city, such as water, gas, electricity, the management of water itself: drainage. It stands for any engineering system availing itself of some aspects of nature, which is used to make a city work. Each of them fulfills an important function for sustaining city life. Since they articulate human concerns and natural forces they can become tools in our hands to differentiate various sectors of the city according to its natural specificities. Until now, city managers have always considered that systems of electricity, gas, water have to be centralized, that all urban systems have to be centralized. These are all very large systems. But in fact recent technologies show that we can both deal with water distribution and electricity in much smaller units, which allow us to create much greater heterogeneity within the cities, servicing local population needs in a better way. So I think we are moving a way from the concepts of 19th century engineering infrastructure that provided a conceptual basis for engineers to claim it was the only efficient way of doing things, and instead we are breaking down each one of them into different shaped systems, responsive to different ecological situations. This new "infrastructure" strives to make different sections of the city work in their own way, rather than the whole city in one and the same way.

A linear park is a part of the city infrastructure in the sense that it creates animal and plant life supporting corridors; therefore it fulfills an important ecological function, which had been wrecked by the enforced separation of city life and nature. And at the same time it gives us the opportunities for the aesthetic creation of a different kind of park.

So the critical examination of infrastructure creates a sort of platform for us to conceive of the old pieces in a very different way. So we can transform the infrastructure and the idea of a park at the same time. Because of what we know about the new technologies available to create urban infrastructure by adapting in a new way to local specificities of nature in the city. We can look at a new piece of infrastructure and say "Oh, a park!" That means it could be something totally different from any infrastructure or any park we know. We transform the concept itself of a park. In the US, and many other places in the world, the park essentially evokes an Olmstedian image that we have fixed in our heads and we take for granted.① But this thing, sitting outside the windows, the High Line②, is a good example of another new kind of park.

---

① Diana Balmori alludes to the park systems created by Frederick Law Olmsted, his firm and later his sons, which are the ultimate reference of picturesque city park in the US. Central Park (1857–1873) and Prospect Park in Brooklyn (1865–1873) in New York City contributed enormously to the development of the popular image of a city park in New York City and in the US.

② The "High Line" is a recent renovation of an abandoned railway running on a bridge in the South of Manhattan, just outside Diana Balmori's office, which has been transformed into a planted mall by James Corner (Landscape Architect) and Diller Scolfidio+Rento (architects) between 2005 and 2009, following a celebrated precedent, the "Promenade Plantée" in Paris by Jacques Vergely (landscape architect) and Philippe Matthieu (architect) between 1987 and 2000.

And, if you go another block, you would see the Hudson River Park [1] and that is still another kind of linear park. So the infrastructure opens our doors to take the old concept and just transform them.

WU: This shift of conceptual paradigm in urban infrastructure design is very interesting. It seems however to predict a new universal rationality, but I know you have been working with Asian companies in Japan, China and Korea for a long time and you do not ignore cultural differences. So I would like you to say how designers in East Asia should both follow your lead and take into account their own culture?

BALMORI: This is one of the most difficult questions to answer, and you have a very important point. But I do want to insist that the major problems that we are encountering in cities around the world are the same: incredibly rapid population growth, strict impossibility for the old city infrastructure to support this population growth, growing need for public space and for healthy city environment in the face of deteriorating climatic conditions. These problems are new for everybody, not just for Asia or America, and solutions have to be new. We are at such a turning point for the ecology of the whole earth, with incredibly fast climatic change for the worse, which the answers have to be planet wide. So on one level, I say the answers are new for everybody; now, the next step when you get into design is how does it tie with local culture and how does it fit local culture. I think that the only good way is, certainly to be very sensitive to local materials, but also, above all, to be able to work with local people. Urban system design calls for team work, and those teams have to be a way of communicating across cultures. There is absolutely no way that I could culturally understand another place no matter how much I study on the spot, I do need somebody to educate me culturally about the place. I was very fortunate when I was working in Japan to have a colleague who educated me culturally about Japan. And I am, in the

---

[1] The Hudson River Park is a recent development of a succession of parks and projects along a greenway that runs from 59th street to Battery Park in the South. It began in 1998 and many landscape architects and architects have been associated with this development that is still underway.

process of trying to set up something similar in India. I think trans-cultural teams are essential for that, I think there is no other way.

Yet, the problems we are hitting are the same everywhere, and we need solutions that are new for everybody. So, the linear park, for instance, is a new way to conceive public space, which is valid for any city in the world, even though it will take totally different expression everywhere. It will use different plants; it will join different things, serve different cultural purposes. But the concept of creating corridors for plants, people and animals is a very valid and a new one, and therefore it supersedes the Olmsted view, even if this does not mean that you cannot create Olmsted parks in addition as jewels on the necklace of these linear parks.

WU: Since you are the designer of a very important new town near Seoul city in Korea, could you explain to what kind of new urban and landscape forms these ideas have led you and your Korean colleagues?

BALMORI: First of all we won an international competition for a new town. The president of Korea at that time had planed this to be a place where 8 different ministries would be moved out of Seoul, because Seoul, like many other big cities, cannot cope with the enormous influx of immigrants. So, he decided to create a second city reasonably close by, from one and a half to two hours by fast train, and to transfer a part of the governmental institutions as a way of creating an economic basis for the city. As things usually go in Korea, at first six different villages of housing towers were constructed all around. Some infrastructure planning were also done, creating a super grid with highways with some incorporated utilities. So, we use this as our point of departure, introducing other systems of connection, to the land and the river which makes an arc around the site of the city in order to break the conceptual homogeneity it was superimposing. This is not a utopia, but a place built to answer to human concerns and natural conditions in each place differently. It was important to demonstrate that. We worked with a Korean architect to prepare construction documents for the first building. Then an international competition was called for the second building within the project. It was won by another Korean architect, who is producing the construction documents at present. It is temporarily halted right now because the next president felt that he did not want the ministries to leave Seoul. He proposed that instead the city be changed to an academic town and built around university functions. But since it had legally been designated as a government city, the process of changing from one to another has legal implications. And those have not yet been finalized.

To start with designing the master plan for the city we took as our departure point whatever the terrain was like, and the infrastructure that already laid down. Because in this case the city was organized around a series of government institutions, we wanted them to be approachable by the public. We therefore limited the height to 6 stories, but contrary to other urban projects, we wanted both the ground plane and the 6th floor to be a continuous public domain. This meant in particular making the 6th floor of all the ministry buildings and the ramps that reach it into a continuous park, giving access in various places to the eight different ministries. Of course there are other cultural institutions that were separated and elements that make the city within the site. But we mostly wanted to respect the place, the sense of being surrounded by high mountains, and to keep them visible from everywhere. We also insisted on not removing smaller hills out of the town but making them part of the city. This goes against customary practice in Korea, where there are so high mountains all over, that city builders cut down every topographical feature in order to erase any new city on a flat plane. So we kept the two major hills that were on the terrain to be able to make them part of the environment, of the real landscape. Then we kept the water flowing on site, the marshes and the existing rice fields, making it clear that rice fields could have a place in the city, and can be used as part of the city water cleaning system. So we were trying to incorporate them for what they are, for their agricultural value, but at the same time for what new services they can bring to the city, for their ecological value. All of this was designed to make the whole city life a source of pleasurable aesthetic experiences. So the attempt was therefore to allow the villagers to find whatever services they needed in the city within a short distance of their home, to enjoy pleasant public spaces in the city, to gain a sense of proximity of public activities of which they were not separated, and also to enjoy from the very tall buildings where they live wonderful views towards the distant mountain as well as down upon a lush and thriving city with all its buildings under a park. It is a new city and a new landscape. And I think it provides some guidelines for future cities.

WU: I find your attention for research quite unusual among landscape designers. And I would like you to explain what kind of activities you are pursuing in the research lab-the"BalmoriLab"-that you have incorporated in your design office. Could you explain how it engaged in a piece of landscape research on green roofs?

BALMORI: The green roof was started precisely as a research project. We did not have a client; but we thought that if we believed in green roofs, we might look at a particular part in New York City and ask ourselves a few questions: how many green roofs could there be in that area and what the effect would be? What would come out of looking at the aggregate of the impact of green roofs?

So we took an area in Long Island City, across Queensboro Bridge on the other side of the East River in front of Manhattan. We took that area because the industry in that neighborhood uses enormous "pancake" buildings. These three stories buildings have enormous flat roofs. Besides, there are also some towers in it and lots of smaller residential buildings. We just made all the roofs green. As a way of approaching the resulting phenomenon as a whole, we just added up all the surfaces, and it turned out to be half the surface of Prospect Park, the enormous park that Olmsted had done in Brooklyn. I thought it was really surprising, it means if you do this you can create a park without buying any land or doing anything else. Even if you do the minimal greening, the "extensive roof", just putting seeds on top of three inches of soil you would get the benefits of natural vegetation on half the surface of Prospect Park. I found that very revealing. It showed that we had not thought of green roofs in the right way, because we had been thinking of the specific benefits for a particular building. But once you put the aggregate together, you really understood that you have the means to clean the atmosphere, to impact the heat dynamic of the air of the city. Even more, in fact, you have the possibility of creating space that can be used for the inhabitants of each building, and eventually become part of the public realm. Many of those huge industry buildings are colossal discount stores. With the minimum of incentive they could turn their green roofs into public space as green as a park in which they could sell outdoor merchandise or whatever. It just requires taking a new approach to the design and use of space. This was just a piece of research we started in the office. Afterwards, we contacted Long Island city and suggested they might be interested in this study of green roofs. They agreed and invited an educational institute addressing real estate developers that was interested in green roofs. Some of their clients came, one was a television studio, the Silvercup. They said ok, we would like to do some green roofs for our studios, which we did; and then there was a modern furniture manufacturer that asked us to do it for his. We even got some grant money to have the ecological impact of these green roofs studied over a year to see the results of what we really made into solid figures. So we learnt that in fact all the things we said were actually exceeded on the test of the Silvercup roofs. We are presently waiting to get the answers about the smaller roof at the end of this year. Though we do not yet know all the results, it was really a nice piece of research and experiment.

WU: Did these studies affect the building code in New York City?

BALMORI: The only thing that happened—a very good thing—is that now there is an acceptance for green roofs. So now you can get tax deduction for your green roof. So it encourages people to do it. And that in turn has been able to produce a small industry of green roofs. There was one installer and there was one plant person when we became involved, and now there are 7 different systems for building green roofs. And the fellow who has studied the plant business has now become national, and he sells plants for green roofs on the national scale, even in the south. So it has even created business, and it could bring about a huge transformation of the industry.

WU: In other words more and more people are getting involved.

BALMORI: Yes, more and more people are getting involved. In New York, I would say there is enormous interest for it, and it may already concern up to 10% of New York's roofs. I think that in a couple of years, up to 20% of the roofs might become green roofs. Nobody is counting them; I just say that from the calls I receive for advice. Right now anybody who has a private building can build it on his own, because they can consult with people who have already done it. It is already well known enough. But nobody thinks as much about the design of the roof. If you just do a very thin skin of sedum, it is not so much an issue of design. I mean it is a minimal issue of design.

WU: *Redesigning the American Lawn* is a research you did and published many years ago. But I think now it still has particular relevance to the Chinese readership.

Can you talk about it?

BALMORI: It feels like old work, and there had been so many books about the American lawn since then, but ours was the first. I did this study with two colleagues from the school of Forestry and Environmental Study at Yale. It is a piece of work that keeps coming back to me because people rediscover it and want more information. I think the gist of it is this: where the climate does not sustain a lawn you should not do it. And the whole west of the US should not use lawns; it is as simple as that. They do not have the water, and lawns use water in such enormous quantity! The other thing is that the lawns that we have developed are very poor biologically and they have basically decimated biodiversity rather than helped it, because they only use one or two species of grass, and they use a lot of chemicals to make it look perfect. Those chemicals kill all sorts of small creatures within the grass that help maintain the soil. So, grass need to be fed out artificially, and enormous amount of nitrogen are sprayed on them and on the ground. Parts of this nitrogen go down into the water below ground, and eventually infiltrate in rivers and create algae growth which kills fishes in the water. So it has enormous consequences, and in fact the studies in the United States show that the chemicals used in the lawns are much more detrimental to river systems than all of the chemicals used in agriculture. So it is very harmful for the environment as a whole, it misuses the water and it is not good for the health of people, because the sprays on the lawn are dangerous, and skin contact should be totally avoided, but is not always. So we are dealing with a landscape which was created because it looked beautiful, but it was created in England, in a climate that could sustain it, that has frequent small rains. First, in the absence of rain, you should not use the public water system to sustain it. Second, throughout the 18th century, until by the end of 19th century the big chemicals and tool companies started to control it, lawns had an enormous variety of plants growing in them. And among these plants, some like clover, when it dies in mid summer because of the heat, produces nitrogen so you do not have to put it on artificially. Moreover this form of nitrogen is absorbable by plant roots and does not go into the ground water. So the regime of even those lawns that were fitted to the climate, were also much more generous to the rest of the atmosphere. Yet, since the end of the 19th century. We developed this industrial lawn, and we ought to understand how detrimental it is to the public environment. Therefore if we do produce lawns, and

I understand the attraction of the lawn for playing on the soil surface with children and for children's games, we should do so in small quantity. We should use lawns where the climate could sustain them. And when we plant them, we should follow a pre-19th century mode with a mix of plants, containing things like clover that will make them sustainable. So on one level our research calls for a modification of the contemporary industrial lawn, but on the other hand it implies that where it is not sustainable, for lack of rainwater, you should not develop it, because it is bad landscape, it is harmful landscape. So it is an important thing in landscape to understand what each piece of the earth can sustain well. It is more interesting and more beautiful to live in a sustainable landscape with its autonomous life and singular changes, and also it ceases to make the whole world into the same thing everywhere.

WU: Beauty and aesthetics, once more.

BALMORI: Yes, I want to add that people who worked on lawns in England in the 18th century really developed it as an aesthetic object and that is why it attracts everybody. Therefore our job today is to make this new lawn into an aesthetic object as a transformed piece, not a pursuit of the industrial lawn. So we also need to be able to work on redefining its design, so it would be as attractive as the 18th century lawn. But it was artistic works, great parks and gardens, done with these lawns that made them so desirable, and we have to create their counterparts as well.

WU: Thank you for a great illustration of some relationships between aesthetics and ecological thinking, and for showing how your thinking constantly shuttles between large environmental issues and site specific relationships between humans and non humans in nature. (Translated by Chan XU, Proofread by Xianming TU, Xin WU)

贝尔纳·拉素斯
——视觉艺术与景观的诗意

贝尔纳·拉素斯是一位法国艺术家和景观设计师,是2009年联合国教科文组织赞助的国际景观设计师联合会终生成就奖——杰弗里·杰里科爵士金质奖章 (IFLA 2009 Sir Geoffrey Jellicoe Award) 的获得者。该奖项是国际景观设计师联合会所给予景观设计师的最高荣誉,每4年授予一次,旨在对"一位在世的景观设计师,他终生的成就和贡献对社会和环境独一无二的长期影响,并促进了景观设计师职业的发展。"(IFLA)予以表彰。拉素斯是法国最早的动感艺术家之一,成名之后,他在20世纪60年代末转向景观艺术,并一直在这一领域实践、研究和教学至今。他独到的景观思想通过他的作品和在许多欧美大学的讲座正变得越来越有国际影响。他的创作有两重中心:对视觉现象的仔细而客观的研究;对其中的人的态度和评判的关注。这一设计哲学在处理高速公路和人文景观的关系时极为有效。20世纪80年代以来,拉素斯形成了公路景观设计的新思路并帮助法国政府制定了专门的《国家公路景观政策》。他强调对田园风光的尊重是法国文化和现代化的一部分,这也是所有的开发商、设计师和工程师都应有的责任。拉素斯历任巴黎美术学院建筑系绘图教授,创立了巴黎第一大学塑性艺术系,并参与凡尔赛景观学院的建立,在那里教授以现象学的方法来研究人与自然间的多重感性关系。

拉素斯教授在此访谈中反思了他的景观设计作品与20世纪后半叶当代艺术以及当代建筑之间的关系。他解释了作为一个景观设计师,他的创作如何贡献于世界视觉艺术(按法语用法，他称之为塑性艺术，包括雕塑和建筑）。这是非常重要的论述，因为很多景观设计师惧怕称自己是艺术家。拉素斯教授将大胆的艺术手法引入景观设计，并以他最近的作品为例，详述了他的一些主要设计理念。他也提及中国艺术、山水画和园林。他甚至认为在中国艺术传统和他的欧洲现代观点之间有很大的相通性。确实，对中国读者来说，如果我们将其与两个传统美学中的概念相连——"意境"和"大象无形"，他在景观艺术中所追求的"氛围"的概念是不难理解的。

吴欣：拉素斯教授，您是一位著名的艺术家和景观设计师。您最近为COLAS集团在巴黎两座代表性现代建筑的天台上设计了一系列的空中花园，是非常引人注目的艺术作品。为了满足居住于此建筑中的当地居民的需求，您将一幢现代主义建筑，一幢典型的纯白构筑和人工机械，转化为富有隐喻森林和抽象花园的公众景观。由此看来您似乎专注于高端艺术，但是您的第一本书是关于巴黎郊区的乡土花园美学和矿业城市的住宅分配研究[1]。您能否解释一下您是如何对民间景观美学产生兴趣的？

贝尔纳·拉素斯（以下简称拉素斯）：在我还是个年轻的学生时，现代建筑与现代艺术，也就是绘画、版画和雕塑的分离令我黯然。这也是我加入巴黎美术学院的费尔南德·莱热[2]（Fernand Léger）工作室的原因之一。莱热，以及后来的瓦萨雷利[3]（Vasarely），都在试图向建筑方向发展，但都没有成功地渗透到建筑界。两位艺术大师都致力于将现代艺术引入建筑，我记得瓦萨雷利告诉过我，这或许可成为我终生为之奋斗的目标。那时，色彩是将塑性艺术重新融入建筑最简单的方法。阿道夫·路斯[4]（Adolf Loos）是一位颇具影响力的建筑师，曾经宣称建筑内将完全杜绝装饰。因此，现代建筑师完全将建筑装饰抛到一边。他们应该更谨慎一些的。完全抛弃装饰意味着他们淘汰了一些处理建筑的手法。我想莱热是第一位注意到这一点的人，并且这后来也成为我创作的焦点。同时，应当说明的是，我是相当反对艺术市场及其操作的。

---

[1] 见贝尔纳·拉素斯所著的《想像的花园：居民景园师》（巴黎：知识出版社，1977）。
[2] 费尔南德·莱热（Fernand Léger，1881—1955年）是一位法国画家和雕塑家，深陷于关于现代艺术的早期辩论中。他认为绘画必须脱离油画的表现方法，并致力于在世界范围内批量生产现代城市和建筑。另外，在应当遵循的艺术原则上，他与柯布西耶的理念大相径庭。他是20世纪20年代从抽象画派分离出来的法国画家之一。
[3] 维克托·瓦萨雷利（Victor Vasarely，1906—1997年）是一位匈牙利画家，整个艺术生涯都在巴黎度过。国际上公认他是20世纪最重要的艺术家之一，欧普艺术运动的知名领导人。
[4] 阿道夫·路斯（Adolf Loos，1870—1933年）是一位奥地利建筑师。他的论文《装饰与罪恶》反对维也纳分离派的装饰风格，对现代建筑的定义作出了贡献。

皮埃尔·弗兰卡斯特尔[①]（Pierre Francastel）有一次告诉我，为了成为一位艺术家，你可以上学获得文凭，也可以不走设定的寻常出路，自我发展。于是我就为自己设计了课程。我喜欢色彩，但也对其他塑性创造方式感兴趣，比如绘画、体量、阴影，以及表现手法。莱热认为所有这些方法都必须应用于艺术作品的创造，包括人体艺术；但我不能想象，比如建筑如何能还原人类的艺术表现呢？然而，当我创造出第一件互动艺术作品时，我被色彩、光线以及运动之间的关系吸引住了：看到色彩如何为洒落在其上的光线所改变，以及光线本身如何被其反射的色彩所改变。我花了大量精力对此进行研究。这自然而然地将我引入了对通俗美学的探索。1961年，我被文化部派到科西嘉岛（Corsica）为建筑列出标准的色卡表，因为政府官员担心村庄民居使用新工业颜料的色谱会损坏科西嘉民居建筑遗产的风貌。我来到科西嘉，见了很多村长，拜访了村子和村民。所有的科西嘉人都告诉我能用新颜色粉刷建筑外墙他们有多高兴，住房和农屋的新色调照亮了他们的生活。因此我拒绝制定一系列限制性的用色准则。在科西嘉一个村子里拜访时，我偶然看到了一个花园，它不像我以往见过的任何东西。发现这个非比寻常的居民创作引发了我对于寻常的乡土花园中通俗文化创造性的兴趣。它向我揭示了公共住宅建筑设计在很大程度上与通俗文化相左，背离普通人的渴求。由于我要知道现代艺术如何能重新介入建筑，所以我开始在居民创作的领域深入研究。一些居民的创作旨在颠覆现代主义的居住建筑美学。我必须指明，那时在有些地方建筑师甚至强行规定了窗帘的颜色，居民被剥夺了选择房间窗帘颜色的权利。不过有些居民不服从这样的规则，因此我开始记录他们的策略。当然，他们的反抗是有限的，通常表现在窗台或阳台上。我想知道除了在阳台上能观察到的部分，哪里还能发现居民更为自发的创作。所以我将注意力转移到宅院的前花园，因为住宅院的人能够控制的范围大于一个蜗居在公寓的人，他们能改变的

[①] 皮埃尔·弗兰卡斯特尔（Pierre Francastel, 1900—1970年）是一位法国艺术历史学家以及艺术批评家。他是公认的艺术社会学奠基人，他的著作涉及艺术和科技，文艺复兴时期绘画空间的发明、衰败并如何被立体主义取代，因此而广为人知。他是拉素斯父亲的朋友，并且在学生时期担任其导师。

是一个房前花园而不仅仅是一个阳台。我甚至在1967年获得了国家科研经费，用以研究住房宅院中的居民自发创作的美学。我称他们为"景观的栖居者"、"居民景观师"，这些居民改变的范围从街边延伸至花园，再一路到房屋立面。通过对他们的作品系统的研究，我意识到把视觉艺术再次引入建筑实践的重要性。

吴欣：所以，对民间文化的研究导致了您后来对自然与建成环境之间关系的关注？

拉素斯：自然和建成环境之间的关系在20世纪60年代争论颇多。我逐渐对动态艺术[①]产生了兴趣，在不同光线和颜色的条件下赋予平面以动感；但我意识到通过一个小引擎或者风将运动引入艺术装置中（亚历山大·考尔德，Alexander Calder[②]那时已经在这个方向上做出成果）的结果可能是一系列的高度重复。这促使我开始关注自然和景观。我观察到森林中的光线和邻近草地上的光线不同：由于光线从一片叶子反射到另一片，因而衍生出一片暖色的空间；而在草地上，由于光线几乎直接照射到地面上，反射的光也是直截了当的，修饰甚少，景色大有不同。我观察到，自然界的动感远比我通过机械手段把动态引入艺术作品中要丰富得多。这些观察将我的兴趣转向了景观。

自从人们发现了创造更为复杂的动感的方法，我不是很确定现在是否还是这样。无论如何，20世纪60年代的这些观察对于我来说是最重要的。然而，我无意于严格地模仿自然，因为艺术不能复制自然或任何世上现存的事物，不管是自然的还是人工的。这就产生了艺术的重大难题：自然与人工之间的相关性

---

① 动态艺术是任意依赖于运动的影响而存在的艺术。历史学家们将20世纪头10年的一些作品命名为最早的动态艺术，但它真正成气候却是在50年代末和60年代，拉素斯与GRAV（视觉艺术研究团体）关系密切，该团体在50年代末至1968年很活跃，并曾为Victor Vasarely创建的Denise René画廊展出。

② 亚历山大·考尔德（Alexander Calder, 1898—1976年）是美国出生的艺术家，毕业于美国一所工程院校，后来决定成为一位艺术家。他1926年搬到巴黎，会见艺术家，并推崇抽象艺术。他在1933年回到美国，后又于1962年重返法国定居。动态艺术家对他的作品评价很高。

如何？我认为艺术的问题就在于，在人工和自然的距离之间建立一种文化认同。一位艺术家不可能对世界作直观的复制。一旦他意识到无法再造自然，就必须探索自然与其作品之间的鸿沟。我认为艺术在很大程度上关注的是去发现超越直观映像之外的自然与人工之间的临界区间。这才是真正的艺术范畴。一件艺术作品，如何才能既表现真实事物而不陷入具象？这是我目前在研究的问题，因为我意识到我终身的创作都在思考这个问题：表象和现实之间的关系。然而，我们或许注意到，超越表象的艺术与非直接再现现实的艺术并存。当人们非具象地表现现实时，你会发现艺术很具有启发性，这一点是很奇妙的。这也是理解毕加索作品的钥匙。你可以看到毕加索在不断地探索具象与非具象表现现实的沟壑。莱热拉开了两者之间的距离，并通过他的作品始终维持着这个距离。而毕加索则相反，他不断探索不同的距离、不同的非具象表现方式。一旦他建立了一个距离，他似乎就会去寻找不那么真实的但依旧能够唤起现实的艺术作品。在我看来，一件艺术作品中最重要的东西就在于发现周围世界的非具象意象。

吴欣：对于这种非具象性的追求是如何与现代艺术相关联，并滋养您自己的创作的呢？

拉素斯：艺术作品的涵义，以我之见，在于对独立于具象表现与对我们周遭世界的启迪能力之间的盲区进行探索。艺术家们必须创造出富有意象又不直接重现这个世界的作品。艺术能够在何种程度挣脱直白的再现？莱热在这方面的成就高于很多与他同时代的艺术家。因为许多人转向抽象绘画时，就放弃了这个问题。尽管这个问题不时会以有限的方式出现在注重质感的抽象派艺术家的作品中，例如，在放弃图形之后通过肌理来唤起对真实物体的触觉。大多数20世纪的艺术作品产生于试图从不同的塑型艺术创作中分离的境况，以及对各自领域潜能的探索。总的来说，这成了现代艺术的任务。我们可以看到，这种不同创造方式的分离，可能是和科学分析齐头并进的。抽象艺术在这种环境下发展起来，但这已经是过去了。这个时期已经终结，而在这段探索的时期过后，我们可以再次投入对非具象的研究，并且将各种可能的创作方式都用上。这就是我至今一直在努力做的事情，尤其是在 COLAS 集团①的花园中。我在 20 世纪 70 年代或者更早时候，做了"彩色灌木丛"②，当时的问题已经是如何摆脱表象来表现树丛。就是那时，我们添加了原本不属于它的彩色斑块，没有复制其表象，但呈现了它的存在。

弗兰卡斯特尔曾经说过，在近代以前，追求真实世界的直白表述曾是艺术的规则，它导致了摄影的诞生。摄像机的发明旨在达到现实主义画家一直引为艺术理想，但又因为不可能而一直无法成功实现的目标。事实上你可以看到在 20 世纪摄影本身也是在这个方向上不断衍变的，拒绝与现实世界的简单复制混淆。同样的，现在我从事非具象艺术的研究工作。可以很容易在 COLAS 集团 7 层平台上的"碧色剧坛"看到这一点。喷泉确实是普通喷泉的尺寸，但却是由霓虹灯的彩色光线组成。而且我曾经尝试不用岩石的表象做出能够唤起岩石的姿态和图形感觉。我预见无论怎样设计模仿岩石，对于制作艺术作品的趣味来说都太直接了。正是艺术作品和真实之间的差距产生了召唤力。这在我最近的作品里体现得非常明显。这也是与他人、与公众交流的问题。如果艺术作品不具有感召能力，其他人也无法与之产生联系。所以召唤力总是关注点。而它也不应屈服于直接表述的吸引力。

吴欣：在您的早期作品 Guénanges 高中的彩色灌木丛中，您是否已经进行了探索？

拉素斯：是的，那是第一次尝试，但只是一个装饰。对于今天我在这里要说的概念里，那并不能说是一件艺术作品。那不是为了避免具象表述而进行的正式研究，虽然也起到了唤起真实世界的作用。

---

① COLAS 集团是由一个法国公司 Colas SA 创建的国家工程公司。它在超过 40 个国家工作，大多数是公路建设的项目。它的总部坐落于巴黎最近的郊区布洛涅比扬古，拉素斯的空中花园就是建在这里。

② 彩色灌木丛是拉素斯为 1972 年建校的 Guénanges 高中创作的艺术装置。为了支持艺术创作，法国文化部已经规定，所有公共教育建设预算的 1% 必须用于艺术作品。彩色灌木丛是用很多彩色金属球做成的抽象灌木丛装置，中和了人工制造和自然环境。

吴欣：您是如何从这些早期的艺术作品转向景观的？

拉索斯：与其说景观，我们更应该用氛围这个词。这意味着在塑性艺术中应当使用所有可能的创作方式。莱热没有进行氛围的创造，虽然他想要使用所有的设计手段，但他半途而废了。我想我已经从自己对于艺术可能性的探索中超越了莱热，即使这听起来有点自命不凡。莱热从对比的角度思考，而我的思维则进展到了异质多重性的角度；这显然不是同一回事。同时，他有远见地尝试了各种可能的创造方式，却没有将光线考虑在内。就他看来光线仅仅是艺术作品存在必须适应的外在环境。而氛围是一个更加复杂和互动的艺术创作，其中光线也是艺术家的一种手段。然后形式在光线下衍生出运动，并被转化而产生氛围，因此只有氛围才是艺术作品的真实形式。在某种意义上，这氛围是一种综合的形式，将所有的感官体验元素集合起来：触觉、气味、温度、光照、材料质地和颜色。氛围是我们环境体验的形式，通过精妙地利用它所有的组分形成创造性的组合。因此氛围的艺术是景观和花园的艺术。从这个角度看，这就是为什么在我的作品中没有不连续性；我探索了所有这些元素的各个方面。氛围是关于我们在这个世界中栖居的体验艺术；它探索为人类生活创造场所的方式。只有色彩是不够的，只有光照也是不够的，有材料质地还是不够的，我们必须利用所有可能的设计手法来激活人类感知。

吴欣：您具体是如何探索这一艺术的新领域的？

拉索斯：1965年，我做了一个称之为"Un Petit Air Rosé 粉色小气象"的艺术实验。这是我早期关于色彩、金属和着色表面实验的转化和继续。但是这里有另一个想法，即郁金香的介入——将纸张与郁金香之间的空间本身视为一种蔓延的物质。这种无形的物质就是氛围。它是一个整体：外光，郁金香自身的颜色，还有与外光相关的空气的颜色，以及它在郁金香花瓣上的反射光。这样我们就可以理解它已经是一种氛围，因为这里有一个交互作用的复杂系统。白色的纸张凸显这种氛围，令其可视，因为大部分人并不理解空气的颜色是由郁金香花瓣营造出的特殊氛围；白纸的介入给人们"欣赏郁金香"的行为以全新的体验。这也说明，艺术作品能够呈现出光的存在：艺术作品不是必然被动地接受光的照射，相反，光也是可以被发掘和塑造的。显然，揭示为空气着色的彩色光线是呈现氛围的一种方式。使用白纸这一最小的介入，将自然的郁金香转化成一件艺术品，一种氛围的表达。

这导致了我在几年以后将极少介入的手法介绍到景观设计中，因为这次试验表明了非常轻微的介入就可以凸显出之前未被关注的氛围，或者也可以说，使得某种被忽略的景观品质获得美学欣赏。它建立在感受和认知两者的区别之上。最低限度的介入使人意识到某种景观或氛围所特有的感性。它使得我们能够感知他们，因为感知意味着意识到这种感受体验。氛围的艺术是为了能够让大众感知到这种交互作用的复合系统，它能够触发与环境作整体的感性交流：光照、物质、颜色、运动……

吴欣：在设计COLAS集团空中花园时，您是如何应用这些想法的？

拉索斯：那时候COLAS集团已经成功地建了两座现代建筑，每一座都拥有一些阶梯状平台，而我的任务是把它们变成花园。当时，他们告诉我最低的平台要用来召开鸡尾酒会或者舞会，而其他的平台禁止员工和访客入内。同时甲方还要求设计的花园无须维护。除此之外，只有一些技术限制。首先让我来介绍最低的平台上四季园的树丛。这是一些很小的树，不足1m高，由一种金属板剪切而成，但树叶有着真正树叶的尺寸，还能看出它们大概的颜色和轮廓。因此每个人都认为它们是树，即使并非如此，它们甚至看起来也不那么像树。我认为这个项目值得重视，因为橡树和它的叶子在尺寸上形成的内在反差是唯一的提示，然而人们却意识到了是橡树。这证实了前面关于召唤力和非具象的讨论。更有甚者，每一棵金属树都为四季而设计成4种颜色，这样它们就可以在任何时间被放在任意位置，这显示出一种基于现实的自由。在这个花园里树叶的颜色可以任意修饰改变，尽管在自然中它们必须随季节的节奏而变化。因此，通过将它们从季节交替的限制中释放出来，我引入了一种新的审美可能。这些树为使用者所爱，他们对树的喜爱甚至超越了我的期望。从平台周边的围栏望出去可以看到很多正常的树,我原本期望,做出与这些真树相较不成比例的扁平树会受到批评。但是这种内在的反差抓住了公众的注意力。而且他们对于这些平树的讽刺寓意理解得非常到位,这个小花园空间是用来开鸡尾酒会和舞会的。因此这片场地必须铺上地板,这样就不可能像通常的花园那样种上树。这个空间将是个花园,它需要树,同时又必须能够跳舞。解决的方法是通过艺术作品创造出一种花园的氛围。然而如何创造这样一种气氛,是一个大挑战。小瀑布在这方面功不可没,因为它的流水的声音充溢整个花园,制造出一种愉悦的氛围。树是平的,而声音淡化了树的体量。水的流动转移了注意力,令人想起自然中的瀑布。小瀑布使场地各处的角落交织起来,成为一个感受空间,将各种各样感知汇总成一个整体的氛围。"粉色小气象"郁金香中的白色纸片呈现出我们通常认为是虚空的光线体量,小瀑布的声音使单看起来只是一片空荡的舞台现出了花园的体量。

最近,我一直在研究具象和意象的问题。我以前从未对此写过只言片语。准确来说这不是新想法了,但最近的一些反映令我开始思考并理解了至今为止我所做的所有事情。乔治·布拉克(Georges Braque)有句话说得好:"只有当创意消逝不见的时候,艺术作品才真正存在。"不然的话,它不是一件作品,而只是一个想法;这不是一回事。无论如何,如果你带着这个非具象的概念来看艺术作品,你会意识到这对于理解当代艺术有多大的帮助。幸运的是,我现在总算是理解了我为COLAS所做的东西,知道了为什么我总是在把玩抽象和实体。

吴欣:"实体"听起来是您作品中的一个重要部分。事实上,在您的整个职业生涯中,您坚持景观中细微差别的重要性,您甚至说过景观是关于微小的坡度的艺术。能否解释一下这个观点?

拉素斯:你的问题提在点子上了,这个观点是很有可能被误解的。讨论微小的坡度时,人们不会强调任何一个感官元素。这与氛围有关。埃菲尔铁塔不是一种氛围,它是一个物体。当然,它是个有趣的物体,但这不是我所关注的。当提及微坡(faint slope)时,我想到的是草地、材质、物质、斜坡所有这些东西。这是一个融洽的整体、一个围合场所、一种氛围。垂直的岩石仅仅是——垂直。

你可以回到我刚刚说过关于抽象艺术的话题。抽象艺术较之尼古拉·普桑（Nicolas Poussin）的画就像悬崖峭壁较之小坡。一幅普桑的画，就像一个小斜坡，首要呈现的是一个相互关系的系统，而抽象画或是悬崖不用。色彩为光、为运动而改变，而光线本身也被改变了，因此它们互相作用（进而改变了我们的场所体验）；但一个物体、一座纪念碑、一面悬崖是不会为相互作用所改变的。随着相对缓慢或加快的跳动，氛围可以说是自然中运动的归纳概括。氛围是自然的，各种缓慢和加速跃动之间愉悦而从容的对话。氛围与纪念碑相反。

2010年春天我可能会被邀请去苏州，因此我开始阅读关于苏州园林的书。非常奇妙的是我的作品和这些地方何其相似。我认为它们是杰出的园林，并非常着迷。15年前，当我第一次也是唯一一次访问北京时，不幸没有时间去看一趟苏州园林。现在我已经心无旁骛地在书中对它们进行研究了，非常渴望能亲身经历。不过，我很高兴在设计COLAS花园以前我没有看过它们，否则我会觉得是在模仿苏州园林。它们看起来与我的作品迥异，但在注意营造意境、避免具象却保留意境方面，这些园林和我的作品异曲同工。当然，我只是从塑型艺术的角度来理解这些花园。每一个花园都是一个水、光线、窗户、起伏的墙面、岩石和亭子互相作用的复合体，没有哪一个造景元素是单独凌驾于其他之上，或者纯粹将其他元素作为背景而脱离出来的。很可能其使用、营造和配植方面的处理都和我的作品不同，不过在更深层次的美学问题上，我是完全认同的。毫无疑问，这些园林具有特殊的氛围，而这些独特的意境使它们成为艺术品。我有幸参观了北京的紫禁城。同样我还是要说，那不是一座纪念碑，而是一种非凡的氛围，一个由大空间和小空间、室内和室外的关系组成的庞大空间系统，湮没了其中纪念碑式的建筑。实际上那里没有空白，节奏非常紧密，以至于整个空间，不管是有无建筑，都像一个整体。紫禁城中的任意一座建筑摆在任何另外一个地方都将沦为一个纪念碑式的大堂。正式建筑和高于它们、围绕它们的院落之间的关系赋予其与众不同的光环。整座紫禁城就是一件杰出的作品。和苏州园林一样，紫禁城也是一个氛围，但两者完全不同。它们如此的不同，又如此同样成功地创造出了有着独一无二紧密感的氛围，这令我感到诧异。

我的作品也是如此。这种不同类似于圣米歇尔山[①]和埃菲尔铁塔。米佑高架桥[②]是另外一个例子。它是一座壮观的桥梁和杰出的工程作品；但它的成功之处大多不是因为大众对工程技术的热爱，而更可能是在于其位于一个美丽景观中，并与之相得益彰成为一个整体。桥的位置非常特殊，而拱形高架桥和场地之间的互动使之成为一个伟大的场所，放射出独特的张力，一种独一无二的氛围。这种氛围超越了桥梁本身。就像中国的紫禁城或长城的氛围远远超越了赋予其名的纪念碑式建筑实体的形象。

吴欣：20世纪60年代您非常关注人类对于地球未知感的消失，这一概念是如何影响您所提倡的关于"不可测量性"的艺术的？

拉索斯：这种未知感毫无疑问正在磨逝。每一代人都体验着未知感，一种要跨越从已知到未知之间的距离的渴望。在现代社会，对每一代人来说这种距离正在快速地消减。这是人类体验的基本特征：遭遇非凡是一种令人敬畏的人类体验，尽管它可能是压迫性的，对于自身居住的世界人类一再地挑战其未知的底线。现今，很多人将我们的文明视作对自然的征用或征服。我认为这是错误的。我们的文明毫无顾忌地驱走了未知的限制；这是人类渴望安全的结果，因为大部分人面对未知世界惶恐不安。而宗教和科学曾经是人类寻求安全感的主要渠道。我对于再次引入面对未知的诗意体验很感兴趣，并不是真正的未知和危险体验，而是其诗情的版本。这也是我捍卫"Heterodite"概念的原因，因为异质多重性在逐渐消逝。这让我能够重塑一种对不安全感的诗意体验。在一个看来非常熟悉、可以完全定位和测量的地方，我引入了一个不可捉摸的、不可测量的异类。也就是

[①] 圣米歇尔山（Mount-Saint-Michel），位于诺曼底和布列塔尼之间一片宽阔而平静的海湾，是建于潮间带岩石岛顶部的一座修道院。12世纪，石头上曾建了一座罗马式的修道院，13世纪重建为哥特式，自中世纪以来成为朝拜的圣地，现代又发展成为法国最有名的风景区。

[②] 米佑高架桥（Millau Viaduct）是世界上最高的公路桥，由结构工程师米歇尔·维洛热（Michel Virlogeux）和英国建筑师诺曼·福斯特（Norman Foster）设计，贝尔纳·拉索斯是景观设计方面的政府指导顾问。该高架桥于2004年落成。

说，在一个完全安全、所有的事情都符合常规的存在中，我加入一个处于所有成见之外的、诗意的事件或尺寸。

吴欣：什么是 Heterodite？

拉素斯：Heterodite 是我对一种景观创造方法的命名；与其将景观看作一个组成部分消失了的平衡整体，这一手法强调景观的异质多重性。它意在揭示人类使用中的历史断点和场地的重要性；与其掩饰交界面，这一手法将不同景观层面之间的断点转化为好奇心的源泉。在某种程度上，我对"Heterodite"的守卫可以归纳为一种单一的呼吁："别再把这世界上仅存的一点异质多重性碾碎了！"因此，人类最有效的景观保护行为应该是尽最大的努力去维护风景的异质多重性，而不是试图将所有的不同纳入到一个统一的模子里。过去是另外一种文化、另外一个国家，是未知的一种形式。

吴欣：这听起来是一个有趣的艺术创意，不过我不明白它怎样能应用到实际的景观设计项目中。

拉素斯：让我用例子来说明。许多年前，我用在兰德斯森林①中拍摄的一套松树的照片制作了一个装置艺术作品。我在这片人工森林里住了一个月的时间；林中只有一种松树，是18世纪时沿着大西洋海岸在绵延数百公里的平地上按照规整的网格排列种植起来的。由于景观的同一性，从中穿越的公路被视作是无聊之极，因为众所周知所有的树都是一样的。那一个月的时间里，我每天都花时间拍摄这些树。对于每棵树，我总是以同样的方式、每天在同样的时间，站在距树皮同样远的地方，从南、东、西、北4个方向分别拍摄。因此，每一张照片上都有大小和比例相同的树皮。然后我把所有这些照片边挨边地排在一起展出。结果显示，所有这些照片每一张都是那么不同，以至于我们难以将同一棵树的4

---

① 兰德斯森林，Landes Forest，在沼泽地上大规模种植10000m²松树（南欧海松）之前，这个地区仍是湿地，并且非常贫乏。它位于法国西南部，并且在大西洋海岸形成了一个巨大的三角地。

张相片重组在一起。这个摄影装置作品发掘了这些树之间或本身不为人知的差异性。它暴露出我们认知的机械性。该摄影作品将未知感重新引入到想当然的熟知的事物中。听到人们谈论松树时我很不悦：似乎它们仅仅是一丛众所周知的复制品；似乎这片松林是一个完全无差别的场所，没有给存在的差异性和生命的创造性，以及令人惊讶的不期然事件，留下一丁点儿可能性。当看着一片人工松林，或者一片宽阔的稻田时，人们可以简单地视之为一个概念化统一的空间，一个直觉可辨并且可用经济计算的实体；或者一个由人类居住的复杂的大千世界，人们在那里会随着自然的波动遭遇各种未知的变化，并不得不适应于变化的无常。如果你要设计一条小径，或是一条铁路，甚至一条高压电线穿过去，你必须先从这两种视角中二选其一。你必须选择是设计为一成不变的空间呢，还是一个出乎意料的差异性空间；你必须选择是将场地简化成众人熟知的地理概念，还是当做不期然的差异性的神秘宝库。作为一位艺术家，如果你对场地的诠释不同，那么你的设计当然也是截然不同的。

吴欣：您可否用 COLAS 集团花园的例子说明一下这不可测量的诗意？

拉素斯：好的，当然。第一件要思考的事情就是金属树丛，放置它们的地方也被命名为"四季园"。这些彩色的树，却没有特定的颜色，因为在花园每处安放点上分别有 4 棵特定的树。花园中的安放点放置有一个特定季节颜色的树，其他 3 棵对应不同季节颜色的树则会收藏起来。并且可以在任何时间替换花园中的那种颜色。自相矛盾的是，这才是让它们成为真树的原因。如果它们被固定下来，就不会特别有意思了。接着解释不可测量的例子：每棵树都是由电脑激光切割的油漆金属薄片组成。效果不可能比得上真树的丰富多样，因为一个两米高涂着红色油漆的金属薄片挂着几片切割出来的叶子，经不起太多的凝视。正是它们几乎不像树的事实使它们可以立在那里成为代表自然的物体，但它们所代表的季节可以由花园所有者随意更换的事实也是迷惑的来源，是认知世界的一种虚假和错位。它们和现行的自然一样不可预测。对花园所有者来说这是个游戏。夏天的时候，他会选择众望所归的夏季叶子、花季的春天、硕果累累的秋天，或者光秃秃的冬日苍穹？他就像四季之神。的确，他是一位超级神明，因为四季甚至可以同时出现，不止有4棵树。可以将选择的决定权留给客人，并且神秘的季节所带来的惊喜超过了大家对一棵真树的期盼。这是设计中引入不可测量的例子，而且我应该说这是那里的访客高度赞扬该花园的主要原因之一。

吴欣：第八层有哪些树呢？哪些花朵和果实？是否也是一些类似的介入？

拉素斯：不，这是另一回事。第八层的这些树在候望台花园里，俯视碧色剧坛的平台。它们的存在大多是为了创造触觉和纯视觉之间的冲突，因为碧色剧坛上的物体和照片一样是平的。这些物体完全由平板组成，上面带有切割轮廓线，并油漆成单调的色彩，这样就从纯视觉角度制造出体量的感受，就像地平线上的山一样不可触摸。与之相反的是，这些树都能结出立体的花朵和果实。对，你可以随意改变它们，或者变换数量。这个游戏强调这些树可触及的尺度，与碧色剧坛的花园形成反差。碧色剧坛是一个空洞的空间，因为禁止入内。而它本身坐落于一个空旷的天台上。触觉和纯视觉的反差凸显了碧色剧坛的非人性化，当然了，这里面有一种反讽的姿态。

吴欣：您对于超越表象而呈现真实的生命存在，以及激发观者的想象力的重要性的兴趣使我想到中国水墨山水中类似的关注。作为前巴黎美术学院教授，您能否与我们分享一些关于中国山水画的高见？

拉素斯：我对中国山水画所知有限。但是其中曲径通幽的手法吸引了我的注意。当看着一幅中国山水画时，我感觉不像西方景观绘画那样面对着一个布景，而是被邀请进入到一连串的相关空间中去漫步。这在毕加索的画中是不可能的，因为它是一个物体，而中国画是一次漫步。你不可能在毕加索的画中去漫步。这就是说，在中国画中，对于画家来说看不到的东西和一眼看到的画面同等重要。这不是毕加索，甚至莱热的画意。对于他们来说，画中展示的东西很重要，而且只有那些东西；没有任何画面之外的东西需要你去探索和关注。从某种程度上来说，我认为西方现代绘画注重的是表面，而中国绘画呈现的是空间。通常画卷中都会有一些人物邀请你追随他们的足迹一同游山玩水。观者接受到这延至画外的邀请,如同身临其境。这种陶醉将平面转换为立体,以其虚空以及个中美景待人索源。毕加索画中的物体是不诱人的,不可渗透有如岩石。中国画是轻灵的,它邀请你的视线上翱翔于云端、下探幽谷深处。而莱热的画强调事物的重量和物质的触感。当然并不总是这样,我应该小心言论。例如,我记得曾在普林斯顿大学的美术馆看到一幅雄奇的中国山水,令我深深着迷,因为它传达出一种很深的触觉感受,而这对我而言是中国画完全不同寻常的一面。(李云圣 译,吴欣 校)

Bernard LASSUS
—between visual art and landscape poetics

Bernard Lassus is a French artist and landscape designer who has been awarded the Sir Geoffrey Jellicoe Gold Medal 2009 by the International Federation of Landscape Architects (IFLA) under the auspices of UNESCO. The IFLA Sir Geoffrey Jellicoe Gold Medal is the highest honor that the International Federation of Landscape Architects may bestow upon a landscape architect. It is attributed once every four years. The medal recognizes "a living landscape architect whose lifetime achievements and contributions have had a unique and lasting impact on the welfare of society and the environment, and the promotion of the profession of landscape architecture." (IFLA)

After gaining early recognition as one of the first Kinetist artists in France, Lassus turned towards landscape in late-1960s and has been practising, researching and teaching in this field since. His approach to landscape design has grown influential, both through his own works and through his teaching in various universities in Europe and the US. A two-pronged concern: for a careful and objective study of visual phenomena, and for attitudes and judgments passed upon these phenomena by people directly exposed to them are the hallmarks of his work. Such design philosophy is proven to be extremely efficient in dealing with relationships between motorways and cultural landscape. Since the 1980s, Lassus has formed a new approach for highway landscape and helped the French administration to develop a *National Landscape Policy for Motorways*. He stresses that the respect to vernacular landscape is part of French culture and modernity, and it is the responsibility of every developer, designer and engineer. Meanwhile, Lassus pursued a career as a professor of drawing at the School of Architecture at the Beaux-Arts in Paris, initiated the Department of Plastic Arts at the Université de Paris 1 at the Sorbonne, and took part in the creation of the Landscape School at Versailles where he taught a phenomenological approach to multi-sensorial relationships between man and nature.

In this interview Professor Bernard Lassus reflects upon the relationships between his work and the development of modern art and modern architecture in the second half of the 20th century. He explains how his work as a landscape designer is a contribution to the world of the visual arts (following French usage he calls it the world of plastic art to include sculpture and architecture). This is an important claim since many landscape architects are often afraid of asserting themselves as artists. Professor Lassus proposes his daring artistic approach to landscape design, and explains some of his main concepts with references to his most recent works. Professor Lassus also remarked about Chinese art, landscape painting and garden making. He even suggests a deep proximity between Chinese art tradition and his own European modern attitudes. Truly, to Chinese readers, the concept of ambience that he has strived to achieve through his art of landscapes may be better understood with reference to two ideas in traditional Chinese aesthetics—"yi-jing" and "da-xiang-wu-xing."

Xin WU(WU hereafter): Professor Lassus you are well known as an artist and a landscape designer. Your recent creation of a series of suspended gardens on the terraces of two buildings highly representative of modern architecture in Paris, for the Colas Group, is a striking art work. To satisfy the demands of the local inhabitants living in front of these buildings you have transformed a modernist building, an example of white architectural purity and mechanical artificiality into a public landscape with metaphorical forests and abstract gardens. Thus you seem to be devoted to the pursuit of high art, but your first book was a study of the aesthetics of popular gardens in the Paris suburbs and housing allotments of the mining cities [1]. Can you explain how you became interested in popular gardens?

Bernard LASSUS (LASSUS hereafter): As a young student I was distressed by the divorce between Modern Architecture and Modern Arts, that is painting, engraving and sculpture. This is one of the reasons why I registered at Fernand Léger's workshop at the Beaux Arts [2]. Léger, as well as Vasarely [3] later, developed his work in direction of architecture, but did not succeed in penetrating the architecture world. Neither Léger

---

[1] See Bernard Lassus. Jardins Imaginaires, Les Habitants-Paysagistes Paris:-Presses de la Connaissance, 1977.

[2] Fernand Léger (1881-1955) was a French painter and sculptor deeply engaged in the early debates around modernist art. He thought that painting had to move away from canvas representation and contribute to the production of the modern city world together with architecture. However he disagreed with Le Corbusier on the artistic path to be followed. He is one of the French painters who broke away with abstract painting in the 1920s.

[3] Victor Vasarely (1906-1997) is a Hungarian artist who worked in Paris during all his artist life. He is internationally recognized as one of the important artists of the 20th century, as the acknowledged leader of the Op Art movement.

nor Vasarely achieved great success in introducing modern art into architecture, and I remember Vasarely once telling me that it would be my life's work. At that time color was the easiest means for re-inserting the plastic arts into architecture. An influential architect, Adolf Loos①, had announced that decoration was to be completely forbidden in architecture. Thus architectural decoration was brushed aside by Modern Architects. They should have been more careful. Under the pretext of getting rid of architectural decoration, they eliminated a number of the means at the disposal of architecture. I think that Léger was the first to notice it, and this later became a focus of my work. Besides, I should say that I was rather opposed to the art market and its practices.

Pierre Francastel② had told me that to become an artist one could either attend a school to receive a diploma, or create one's personal course of education outside the well trodden paths. So I devised my own course of education. I was interested in color, but also in other means of plastic creation such as drawing, volume, shadow and also to a lesser degree by issues of representation. Léger thought that all these means had to be used in order to create a work of art including representation of the human body, but I could not see, for instance, how architecture might re-instate human representation. However as I developed my first art interventions I became attracted to relationships between color, light and motion; seeing how color was modified by light falling onto it, and also how light itself was modified by the color upon which it was reflected. I devoted much work to these issues. This led me to a serendipitous discovery of popular aesthetics. In 1961, I was sent by the Ministry of Culture in Corsica to define the list of legitimate colors to be used on buildings, since government officials were concerned that the introduction of new industrial color pigments on village buildings was about to deface the heritage of Corsican vernacular architecture. I went to Corsica and I met many village Mayors and visited villages and their inhabitants. All Corsican people told me how glad they were that new colors were available for painting buildings

---

① Adolf Loos (1870 -1933) was an Austrian architect. His essay Ornament and Crime against the decorative style of the Vienna Secession contributed to the definition of modernist architecture.

② Pierre Francastel (1900-1970) was a French art historian and art critique. He is considered a founder of the sociology of art, and well known for his books on art and technology, and on the invention of pictorial space in the Renaissance and its destruction and replacement by Cubist art. He was a friend of Lassus' father and a mentor to Bernard Lassus during his student years.

outdoors since the new colors on their houses and farm buildings brightened their lives. So I refused to produce a set of limiting color rules. During a visit at a village in Corsica, I came by chance upon a garden installation that did not look like anything I had seen before. This discovery of a very unusual popular creation triggered an interest for the creative aspect of popular culture that everybody could see in vernacular gardens. It revealed for me the extent to which the architecture of mass housing had failed to engage with this popular culture and respond to its cravings. Since I wanted to know how modern art could be re-introduced into architecture I pursued my research on this domain of popular creation. Some of these popular creations aimed at subverting the modern mass housing architectural aesthetics. I must say that at this time, there were places where architects even imposed rules for the color of curtains, and inhabitants were denied the right to choose the color of curtains in their rooms. Nevertheless there were some inhabitants who resisted these rules, and I began to make note of their strategies. Of course their interventions were limited, usually to the window sill or the balcony. I wondered where I could discover more popular creations than the ones I could observe on balconies. So I turned my attention to front gardens in housing allotments, since single family home dwellers had control over a larger area than apartment dwellers, a front garden instead of a mere balcony. I even received a research grant in 1967 to study the aesthetics of popular interventions in housing allotments. I have called "Landscape Dwellers," "Habitants Paysagistes," the people who had transformed the domain extending from the street into the front garden and all the way to the house façade. The systematic study of their work enabled me to understand how important it would be to re-introduce the visual arts into architectural practice.

WU: So the study of popular culture paved the way to your later attention for relationships between nature and the built environment?

LASSUS: Relationships between nature and the built environment were much debated in the 1960s. I had developed an interest for kinetic art [1] —the transformation of

---

[1] Kinetic art is any art relying on motion for its effect. Historians name a few works of art from the 1910s as the first of the kind, but it really became an established movement in the late 1950s and 1960s. Lassus was closely associated with the GRAV (Groupe de Recherches d'Art Visuel) which was active in the late 50s until 1968 , and exhibited at the galery Denise René created by Victor Vasarely.

a surface in motion under varying conditions of light and color— but I realized that the motion introduced in an art installation by the use of a small engine or the wind (Alexander Calder[①] had started presenting some works in that direction) led to a highly repetitive set of possibilities. This is what drove my attention for nature and landscape. I observed that light in a forest was not the same as in the neighboring meadow, because of the reflection of light bouncing from one leaf to the next, giving rise to a volume of warm colored light, quite different from the light in the meadow, which was hardly modified by reflection since light was reaching almost directly the ground. I observed then that the motion of nature was far superior to the movement I could introduce in an art work with mechanical means. These observations led me to an interest in landscape.

I am not certain that this is still true at present, since one has discovered means of creating motion which allow a greater complexity. In any case these observations made in the 60s have been most important for me. I did not however engage in the strict imitation of nature, since art cannot reproduce nature or anything in the existing world whether a natural or artificial object. This raises one of the great problems of art: what is the relevant relationship between nature and artifice? I think that the problem for art is to establish a culturally legitimate distance between an object and nature. An artist cannot produce a literal rendition of the world. As soon as he realizes that he cannot reproduce nature, he has to explore the gap that exists between his work and nature.I think that art is to a large extent concerned with the discovery of the liminal world beyond literal evocation of real objects. This is really the domain of art. How can an artwork be evocative of real things, without proposing a literal representation? This is something on which I am pondering at present, since I realize that all my life work has been devoted to this question: the relationship between appearance and reality. However, one may observe that beyond issues of appearance art is confronted with issues of creation as a non literal rendering of reality. It is quite fascinating to observe how art can be evocative when avoiding literal representation. This gives a key to understanding Picasso's work. One can see how Picasso is constantly exploring the gap between literal and non-literal renderings of reality. Contrary to Léger who has invented a distance between the

---

① Alexander Calder, (1898-1976) American born artist who graduated from an engineering school in the US and later decided to be an artist. He moved to Paris in 1926, met artists and embraced abstract art. He returned in the US in 1933, but came back to live in France in 1962. Kinetic artists showed great interest in his work.

two, and has maintained the same distance throughout his work, Picasso has explored different distances, different ways of being non-literal. Once he has established a distance, he seems to be looking for what might be even less literal and still remain an artistic work capable of evoking reality. In an art work it seems to me that the most important is the discovery of non literal evocations of the world around us.

WU: How does the quest for non-literality relates to the concerns of modern art and informs your own creations?

LASSUS: Art, as I understand it, explores the domain of artificial works between detachment from literal representation, and capacity of evocation of the world around us. Artists have to create evocative works without attempting literal reproduction of the world. How far can art strive away from literal reproduction? Léger's achievement in that respect is the more important since many of his contemporary, who engaged in abstract painting, had abandoned this issue. It has resurfaced, albeit in a limited way, in works by abstract artists who were creating highly textures surfaces. They evoked the tactile texture of reality, after eliminating drawing for instance. Most 20th century art endeavors have been conducted through a dissociation of the different plastic means of creation and the exploration of the potentialities offered by each of them. In a general sense, this has been the task of modern art. We can see how this dissociation of the means of creation, may have paralleled processes of scientific analysis. Abstract art developed in that context, but this is already the past. This period has come to an end, and we can now, after this period of exploration, again engage in a research of non-literality making use of all possible means together. This is really what I have been endeavoring in the Gardens of the Colas Group [1] in particular. Already with the "colored

---

[1] The International Colas Group is a civil engineering firm created by a French firm, Colas SA. It works in over 40 countries mostly in road construction. The Headquarters, where the Suspended Gardens by Lassus have been built, are located in Boulogne Billancourt, a close suburb of Paris.

[2] The Colored Bushes is an art installation by Lassus for the High School of Guénanges constructed in 1972. The French Ministry of Culture, in order to support artistic creation, had ruled that 1% of the budget of all public education construction had to be devoted to an art work. The Colored Bushes are an installation of abstract bushes made of highly colored metallic spheres which mediates between the artificiality of the construction and the natural surroundings.

bushes[2]", in 1970 or before, my problem was how to move away from the appearance of the tree bush in order to represent it. It was there, and we added color patches that did not belong to it, did not copy its appearance but evoked its presence.

Francastel used to say that the quest for a literal reproduction of the real world, which had been the ruling principle of art before the modern period, had led to the invention of photography. The invention of the camera aimed at achieving what realist painters took as the goal of art and failed to succeed in doing, simply because it was impossible. And in fact one can see that photography itself has been evolving in that direction during the 20th century, refusing to be confused with a literal representation of the real world. In the same way, at present, I am engaged in a research of non literality in art. This can easily be recognized at the "Green Theater" on the seventh floor terrace of the Colas Group. The fountain has the dimensions of an ordinary fountain, even though it is made of colored light in neon tubes, and I have tried to make movements and drawings of rocks that are evocative of a rock while avoiding giving it the recognizable appearance of a rock. I could see that whatever design imitating a rock would be far too literal to achieve an interest as a work of art. It is the distance between art work and the real that lends force to the evocation. This is something that is apparent in many of my recent works. There is also the issue of communication with others, with the public. Unless the art work is evocative others cannot relate to it. So evocation is always a concern. Yet it should not yield to the attraction of literal representation.

WU: Was it an issue you had already explored in your earlier work of the Colored bushes at the High School of Guénanges?

LASSUS: Yes it was a first attempt, but it was a decoration. It was not yet an art work in the sense I am trying to account for here. It was not a formal research avoiding literal representation, and yet achieving evocation of the real world.

WU: How did you evolve from these early art works to landscape?

LASSUS: Rather than landscape one should speak about ambiance. It implies that one should make use of all means of creation in plastic art. Léger did not engage with the creation of ambiance, even though he wanted to make use of all plastic means, he stopped on the way. I think that I have moved beyond Léger in my own exploration of the possibilities of art, even if it sounds a little pretentious to say so. Léger was thinking in terms of contrast, I have developed my thinking in terms of heterogeneity; this is absolutely not the same thing. And at the same time he was right in attending to all possible means of creation, but he had not thought of light as one of them. He thought of light as a given exterior to the means of plastic creation. Light was an outsider to which the art work had to submit. An ambiance is a more complex interactive artwork, within which light is a means in the hands of the artist. Then forms falling under the light engage in motion, they are transformed by this movement which produces the ambiance, so that ambiance is the true form of the work. In a sense the ambiance is a comprehensive form that incorporates into a whole all the elements of sensual experience: tactility, odor, temperature, light, material texture and color. Ambiance is the form of our experience of the environment, and it is achieved by the deliberate use of all its components in a creative way. So the art of ambiance is the art of landscape and garden. This is why, from this perspective, there is no discontinuity in my art work. I have simply explored different aspects of all of this. Ambiance is an art addressing our experiences of dwelling in the world. It explores ways of creating a place for human life. Color is not enough, light is not enough, material texture is not enough, we have to use all possible means of plastic intervention engaging our senses.

WU: Can you explain how you have explored this new domain of art?

LASSUS: In 1965 I made an artistic experiment, which I have called "Un Petit Air Rosé" (a little rosy air). It is the transposition of earlier experiments I had made with color and metal, painted surfaces. But there was another idea, and this is where the tulip intervenes, the idea that the volume that separated the paper from the tulip was itself an expanse of matter. This is what ambiance is about. It was a whole: there was the external light, the tulip own color, and the internal color of the air that related to the color of external light and to its reflections on the petals of the tulip. So we can see that it was already an ambiance, since there was a complex system of interactions. The white paper revealed the ambiance, it made it visible, since most people do not see that the color of the air is an attribute of the specific ambiance created by the corolla of a tulip, it introduced a new understanding of our experience of looking at a tulip. It showed also that a work of art can show the presence of light, that an art work is not necessarily a passive recipient of light, but that light can be revealed and fashioned. Revealing the presence of the colored light tinting the air is, of course, a way of revealing the ambiance. The introduction of the white paper, a minimal intervention, transforms the natural tulip into an artwork, the presentation of an ambiance.

This led me a few years later to the introduction of the idea of minimal intervention in landscape design, since it showed that a very light intervention could reveal an ambiance which had escaped attention until then, or if you like in other words, allow aesthetic appreciation of the qualities hitherto ignored of a landscape. It builds on the difference between sensation and perception. The minimal intervention brings an awareness of the sensations procured by a landscape or an ambiance. It allows us to perceive them, since to perceive means to become aware of a sensual experience. The pursuit of an art of ambiance aims at enabling the public to perceive the complex system of interactions that triggers sensual relationships to the environment as a whole: light, matter, color, motion...

WU: How did you make use of these ideas in the creation of the Suspended Gardens of Colas Group?

LASSUS: The Colas Group had built successively two buildings of modern architecture, each with a few terraces, and I was commissioned to transform them into gardens. However I was told that the lowest terrace had to be used for cocktail or dancing parties, and that none of the others should be accessible to the staff or visitors. I was also requested to make gardens with zero maintenance. Otherwise there were only technical constraints.

First let me turn, to the trees of the Garden of Seasons on the lowest terrace. These are minuscule trees, hardly one meter high, cut out of a metallic slab, but the leaves, however schematic their color and outline, have the size of real tree leaves. So, everybody sees them as trees, even though they are not, and they do not even look like trees. I think that this project deserves attention because the internal contrast which exists between the overall size of an oak tree and its leaf is only evoked and yet people see oak trees. This illustrates our earlier discussion of evocation and non-literality. Moreover the fact that each metallic tree has been designed in four different colors one for each season and that they can be put into place at will at any time demonstrates a freedom with respect to reality. In this garden the color of tree leaves can be modified at will, whereas in nature they have to submit to the rhythm of seasons. So, by liberating them from the constraints of the passing of the seasons I introduce a new aesthetic possibility. Visitors are very much amused by these trees and they enjoy them, even beyond my expectations. I had expected some criticisms for producing flat trees out of proportion with real ones, which can be seen beyond the fence around the terrace. But the internal contrast captured the public's attention. Besides they understood very well that these flat trees offered an ironical commentary to the demand that this small garden space be used for cocktail parties and dancing eventually. So the ground had to be covered with a floor, which would not have been possible if it had been planted with trees as gardens usually are. Yet for this space to be a garden it needed trees, and it was also necessary to be able to dance. The answer is an artwork creating a garden ambiance. Yet, how to create such an ambiance was a big challenge. The cascade plays a considerable role in that respect, because it creates sounds that fill the garden. It produces a musical volume. The trees are flat and the sound replaces the volume of the trees. It displaces attention, bringing an evocation of natural waterfalls, of the movement of water. The cascade weaves together the different parts of this place, makes it into a sensorial space, and knits together the various perceptions into a unique ambiance. As the white piece of paper in the "Petit Air Rosé" tulip revealed a volume of light where we usually see emptiness, the sound of the cascade reveals the volume of a garden, where the gaze alone would have only seen the emptiness of a dancing stage.

Recently, I have been working a lot on these issues of literality and evocation. I have never written about them. These are not exactly new ideas, but rather recent reflections that help me think that I understand what I have done up to now. A word of Braque comes to mind: "It is when the idea of the work of art has vanished away that the work exists." Otherwise it is not a work of art, it is an idea. This is not the same thing. Any way if you start looking at works of art through this concept of non literality you will realize that it brings considerable help in developing a visual understanding of contemporary works. At any rate, now I understand what I am doing for Colas, and why I am always engaging in play with abstraction and concreteness.

WU: Concreteness sounds like an important aspect of your work. In fact, during your whole career you have insisted on the importance of small differences in a landscape, and you have even said that landscape is the art of faint slopes. Can you explain this last idea?

LASSUS: Your question is well taken, this idea could be misunderstood. When discussing faint slopes one does not put stress on one of the sensual features. This is about ambiance. The Eiffel tower is not an ambiance, it is an object. Of course it is an interesting object, but this is not what I am concerned with. When evoking a faint slope I think of grass, texture, matter, slope all together. It is a soft whole, an encompassing place, it is an ambiance. Vertical rocks are just that, vertical. One can repeat about them what I said a moment ago about abstract art. Abstract art compared to a painting by Nicolas Poussin matches a cliff compared to faint slopes. A painting by Poussin, like a faint slope first and foremost presents a system of interrelationships, an abstract painting or a cliff do not. Color is modified by light, by motion, and light itself is modified so that there are interactions (which modify our experience of place), whereas an object, a monument, a cliff are left almost unchanged by interactions. Ambiance can be said to be a generalization of motion in nature, with its relative slow or accelerated beats. Ambiance is the result of the unflinching playful dialog between many slow and accelerated beats of nature. Ambiance is the opposite of monument.

And, for instance, since it is possible that I be invited to go to Suzhou this spring, I have been looking at books about the Suzhou gardens. It is amazing how close my work is to these places. I think they are remarkable gardens. I am quite fascinated. On my first and only visit to Beijing some fifteen years ago I did not, unfortunately, find time to go to the Suzhou gardens. I have intently studied them in books, and I am very much looking forward to visiting them. However, I am glad I did not see them before I designed the Colas garden since I would have felt as if I were imitating them. They do not look like my work, but there are aspects of the attention for creating an ambiance, for avoiding the literality while remaining evocative, which are absolutely common to these gardens and my work. I can only speak of my plastic understanding of these gardens of course. In each of them there is a complex system of interactions between water, light, windows, undulating walls, rocks, pavilions in which there is not a single plastic element that dominate over all others, that detaches itself making all others into a mere background. It is quite likely that the issues of use, construction and cultivation are quite different from the ones I deal with in my work, but nevertheless there are deeper aesthetic issues with which I feel in complete agreement. Undoubtedly these gardens procure special ambiance, and their creation of these unique ambiance is what qualifies them as works of art. I was lucky enough to visit the Forbidden City in Beijing. There again I would say that it is not a monument, but an extraordinary ambiance, a vast system of relationships between large and small spaces, interior and exterior spaces that engulfs the monumental buildings within it. In fact there is no empty space, the air is so dense that the whole space, whether built or not, is like a solid. Any building of the Forbidden City located in another place would just be another monumental hall. It is the relationships between these buildings and the courts that precede and surround them that endow them with an extraordinary aura. The whole is an extraordinary achievement. It is as much an ambiance as a Suzhou garden, but it is a completely different one. It baffles me how different they are in spite of their common success in creating ambiance of a unique intensity.

That is what my work is about. It can be understood with reference to the difference between the Mount Saint Michel [1], and the Eiffel Tower. The Highway Bridge of Millau [2] is another example. It is a spectacular bridge and great work of engineering, but its popular success is not mostly due to popular enthusiasm for technological feats, but rather to its embedding within an extraordinary landscape, of which it has become an integrated part. The location is exceptional and the interaction between the high curving bridge and the site makes it into a Haut Lieu, (a Great Place) which radiates with exceptional intensity a unique ambiance. It transcends the bridge. Just like the Forbidden City or the Great Wall of China transcends the monuments of which they borrow the name.

WU: In the 1960s you expressed great concern about the disappearance of a sense of the unknown on earth. Can you say how it led you to propose the development of an art of the Unmeasuarable?

LASSUS: There is no doubt that the sense of the unknown is waning. Each generation of humans experiences a sense of the unknown, a sense of the distance to be traversed to be confronted with the other of knowledge, the unknown. And in modern times this distance has receded further away for each generation. This is a fundamental feature of human experience:

---

[1] The Mount Saint Michel, is an abbey built on top of a tidal rock island standing in a very large and flat bay in between Normandy and Brittany. A Romanesque monastery was built on top of the rock in the 12th century, and later rebuilt in the gothic style in the 13th century. It was a destination for pilgrims since the Middle Ages and has developed in modern times into one of the most famous landscape sites in France.

[2] The motorway viaduct at Millau is the highest motorway bridge in the world. It was designed by the structural engineer Michel Virlogeux and British architect Norman Foster. Bernard Lassus was the government advisor for landscape design. The viaduct was inaugurated in 2004.

the encounter with the extraordinary is a formidable human experience, but it can be stressful, so humans forever push forward the limits of the unknown aspects of the world they live in. Lots of people nowadays see our civilization as exploitation of or mastery over nature. I think this is a mistaken view. Our civilization relentlessly drives away the limits of the unknown. This is a consequence of human cravings for security, because most humans feel unsecure when confronted to the unknown. Religion and science have offered the main channels towards this sense of security. So, I was interested in re-introducing possible encounters with a poetic experience of the unknown, not a real experience of the unknown, a real danger, but a poetic version of it. This is also why I have battled for the heterodite, since heterogeneity is on the wane. This enables me to re-inscribe the sense of insecurity within a poetic experience. In a place which seems well known, fully mapped and measurable, I introduce an unaccountable difference, something that escapes all known measures. Thus, within a fully secure situation where everything matches ordinary expectations, I introduce a poetical event or dimension that evades all known explanations.

WU: What is Heterodite?

LASSUS: Heterodite is the name I have given to an approach of landscape creation that eschews composition of a landscape as a balanced whole, and instead highlights heterogeneity. It consists in revealing the historical discontinuities of the human use and significance of a site, and rather than disguise the limits it makes the discontinuities between different layers of the landscape into a source of wonder. In a way my defense of heterodite is simply a call: "Do not crush what little heterogeneity is still left in this world!" So the greatest effort at conservation in which humans might engage should be an effort at conserving heterogeneity, rather than keep striving for integration of all differences within a common mould. The past is another culture, another country, a form of the unknown.

WU: This sounds like an intriguing artistic project, but I do not understand how it could apply in a real landscape design situation.

LASSUS: Let me give an example. Several years ago I have created an installation

---

① Until the massive plantation of some 10,000 square kilometers of marshy land with a single species of pine tree (pinus pinaster), this region was extremely poor. It is located in the South West of France and forms a large triangular site all along the Atlantic coast.

of pine tree photographs shot in the Landes forest [1]. I spent one month in this artificial forest which was planted with only one kind of pine trees along a uniform grid over several hundred kilometers of flat land along the Atlantic coast in the 18th century. The road that runs through them is considered as terribly boring because of the landscape homogeneity, because all trees are known to be the same. I spent a month taking photographs of these trees everyday. For each tree I always stood in the same way at a short distance from the bark, at the same hour of the day, and I shot a picture of each tree from the south, east, west and north directions. So, each of these photographs shows a rectangle of bark in exactly the same size and proportions. Then I made a presentation of all these photographs aligned side by side. It reveals that all these images are so different from one another that we cannot reconstruct the groups of four images of a particular tree. The photographic installation reveals the unsuspected differences within and between trees. It reveals the artificiality of our knowledge. The art photography installation re introduces a sense of the unknown in something that was deemed well known. I am unhappy when hearing people speak about pine trees, as if they were just a bunch of well known clones, as if the pine forest were a completely undifferentiated place, leaving no possibility to the existence of differences, to the display of creative life, to unexpected events with which humans are confronted whenever you look at an artificial forest of pine trees, or an expanse of rice fields you may see either one as a conceptually uniform space, an entity which is given to immediate knowledge and amenable to economic calculation, or as a world of differences inhabited by humans who are confronted there to the unexpected variations of the flow of nature and who struggle with its vagaries. If you want to design a footpath, or a railway or even a high power electric line traversing them you have to choose from which of these two perspectives you look at them. You have to choose whether your design would interact with an undifferentiated space, or with a space of unexpected differences; you have to choose whether the site could be reduced to a well known geographical concept, or perceived as a mysterious treasure chest of unwarranted differences. Depending on your interpretation of site, as an artist your design would certainly be different.

WU: Can you give an example of this poetic of the unmeasurable in the Colas Group gardens?

LASSUS: Yes, of course. The first thing that comes to mind is the group of metallic trees, which have given its name to the "Garden of Seasons" in which they are placed. These are colorful trees without a proper color, since there are four identical trees for each location. One, with the color of a particular season, stands in the garden; the other three, with different colors corresponding to other seasons, are in storage. The one in the garden can be changed at any moment. Paradoxically, this is what makes them into real trees. If they were fixed they would not be of particular interest. Let me explain how it is an example of the unmeasurable: Each tree is made of a thin slab of metal cut with a computer guided laser, and then painted. The result cannot match the richness of variations that can be found in a real tree, because a 2 meters high slab of metal covered with red paint and presenting a few leaves cut out of it does not offer much to the gaze. The fact that they are barely looking like trees locates them within the realm of objects representing nature, but the fact that the season they evoke can be changed at will by the garden owner introduces a source of bewilderment, a hollow, a fault in the knowledgeable world. They are as unpredictable as living nature. For the garden owner it is a game. In the summer he may choose the expected summer foliage, the flowery spring, the fruitful autumn or the bare winter canopy? He is like a God of the Seasons. In truth he is a Super God, since there are more than four trees, the four seasons can be present at the same time. It will be left to the guests to make sense of each choice, and the mystery of the presence of the seasons introduces a source of wonder even greater than what the public would lavish on a real tree. This is an example of the introduction of the unmeasurable in a project, and I should say that it is one of the major elements that make this garden highly appreciated by the people who are invited there.

WU: What about the trees on the eighth floor, with their flowers and their fruits? Would you consider this as a similar intervention?

LASSUS: No, this is a different matter. These trees on the eighth floor are in the Waiting Room Garden, overlooking the Green Theater terrace. They are there mostly to create an opposition between tactile and pure visuality, since the objects on the Green Terrace are as flat as pictures. These objects are made of entirely flat slabs with cut out lines and painted flat colors, which offer an experience of volumes discovered through pure visuality, as devoid of tactility as a mountain at the horizon. To the contrary these trees bear three dimensional flowers and fruits. You are right, one can change them at will or simply modify their number. However this game stresses the tactile dimension of these trees, and the contrast with the Green Theater garden. The Green Theater is an empty space, since nobody is allowed to go there. It is itself located on an empty terrace. The contrast between tactile and pure visuality highlights the dehumanization of the Garden Theater, of course this is an ironic gesture.

WU: Your interest for revealing the presence of life beyond representation, and the importance of stimulating the imagination of viewers calls to mind similar concerns in Chinese landscape ink painting. As a former Professor at the Beaux Arts in Paris, can you share some remarks about Chinese landscape painting?

LASSUS: I do not know much about Chinese landscape painting. Yet the way in which it proposes a path of exploration draws my attention. When looking at a Chinese landscape painting I do not feel confronted to a setting as in a Western landscape painting, but rather to an invitation to roam through a succession of places. This is something that does not happen in a painting by Picasso, because it is an object, whereas a Chinese painting is a stroll. One does not go on a stroll in a painting by Picasso. This is to say that in a Chinese landscape painting the unseen is as important as what is given to the direct gaze by the painter. This is not the case in a painting by Picasso, or even by Léger. For them it is what is shown in the painting that matters, and only that. There is nothing beyond the painted surface that demands exploration and commands attention. In a sense I would say that Western modern painting remains a surface, whereas Chinese painting presents a volume. Often there are characters on the scroll who invite you to follow them in their own exploration of the landscape. This extends to the viewer an invitation to immerse himself into the landscape as he roams into it. This immersion transforms the flat surface into a volume with its hollows and its inner landscapes awaiting exploration. Picasso's painting proposes an object as uninviting to penetration as a boulder. Chinese painting is ethereal; it invites a flight of the gaze into the clouds and the recesses of the valleys, whereas Léger's painting asserts the weight of things, the tactility of matter. This is not always true, and I have to be cautious in my remarks. I remember, for instance, seeing, at the Princeton University museum, a magnificent Chinese painting of rocks. I was fascinated because they conveyed a deep impression of tactility, which was entirely unusual for me in a Chinese painting. (Translated by Yunsheng LI, Proofread by Xin WU)

帕特里夏·约翰逊
——公共艺术与环境基础设施

帕特里夏·约翰逊（Patricia Johanson）是一位美国环境艺术家和景观设计师，也是目前新兴的生态设计的先驱之一。在她长达50年的职业生涯中，她始终执著于景观艺术的环境和社会功能，尤其是其在缓解城市问题和改革自然美学方面的作用。在第二次世界大战后的纽约，她作为早期极简主义画家之一，开始了其职业生涯。1969年，来自《家园与花园》（House & Garden）杂志一份始料未及的花园设计委托戏剧性地改变了她的艺术方向。这一委托催生了150个花园设计方案和7篇关于新风格花园的文章。1969年的《家园与花园》方案中所触及的许多问题在当今景观设计界已成为热点话题，如绿色屋顶、废高架铁道和生态基础设施等。尽管没有最终实现，这份委托为该艺术家的创作开启了一个新的篇章，指明了她从此为之奋斗的目标——"将整个世界定义为艺术"。从那时起，她转而专攻景观，形成了一种融合了艺术、景观设计、土木工程、环境与生态的高度个人化混合艺术。20世纪80年代以来，她的作品受到了越来越多的关注。近年来，她正与工程师们合作，设计大型公共基础设施，如废水处理厂、湿地公园和城市步/自行车道系统。这次访谈从《家园与花园》方案开始，但重点将放在她近来的建成项目上。

约翰逊的职业生涯揭示了第二次世界大战后美国艺术一个鲜为人知的发展方向。她推动了景观艺术的复兴，却不羁于众所周知的文化趋势：形式主义、现代主义、大地艺术和环境保护论。最后，约翰逊是一个中国艺术和哲学的忠实爱好者。她将非西方（尤其是中国和美洲印第安）的自然美学和伦理转译为解决当代城市问题的良策。犹如一位优秀的中医，她坚信景观设计的核心在于以通过激活场所、人和文化的内在活力的方式来治愈地球，将断裂的山水脉络重新连接起来是她的艺术目标。

吴欣：约翰逊女士，很多人会从您在加州佩塔卢马（Petaluma）[1]刚建成的项目中得到启示：那是一个将基础设施性的废水处理厂变为休闲性湿地公园的精彩案例。

帕特里夏·约翰逊（以下简称约翰逊）：从许多方面来讲，这个新的废水处理厂的设计是针对一个美国普遍存在的问题给出了一种不同寻常的解决方案，可能会引起广泛的共鸣。佩塔卢马是一座拥有50000人口的小城，位于加利福尼亚州索诺玛县。和美国的许多其他城市一样，它曾有一个超负荷运转的旧供水系统。市议会对饮用水的缺乏、环境质量以及濒危物种等问题感到忧虑；同时也希望政府和公民提高对这些问题的认识。1998年8月我首次受邀进行调研，随后就受委托与卡罗洛工程院一同设计一个新的废水处理厂。这个处理厂刚刚建成，并于2009年7月开始运营。

吴欣：这是一个功能化景观的绝好范例。在中国，尽管现在建了很多新的湿地，人们对湿地在水净化方面的作用关注得还很少。在我们详细讨论这个新项目之前，我想首先了解您的艺术发展过程，从历史的角度来定位它。在20世纪60年代中期的纽约，当您通过《家园与花园》的委托[2]而对花园和景观设计产生

---

[1] 佩塔卢马项目将公共景观艺术、基础污水处理、动物栖息地恢复以及城市农业结合为一体。通过在主要的城市基础设施中融入具有生命的、活的自然，约翰逊创造了一个多用途的公共景观。她利用人工湿地和自然湿地营造了4.8km的公共步道，用以休闲、教育计划、自然研究和旅游。这个公园与价值1.5亿美元、能同时处理城市污水和雨水径流的埃利斯河水处理厂结为一体，不仅为野生动植物创造了栖息地，也为人们提供了一个将生活废水转化为饮用水的可视性演示。有关建成景观的图片见以下网站：http://www.flickr.com/photos/scottthessphoto/sets/72157621028458861/ 和 http://aerialarchives.com/Ellis_Creek_WRF。

[2] 1969年的《家园与花园》杂志委托，促使约翰逊写出了7篇创新风格的花园的文章，并做了150个相应的设计方案："英里长的花园……线型花园"，"视野外的花园……视线尽头的花园"，"自然与反自然……人工花园"，"感觉与思想的花园……虚幻花园"，"滋养、漂移、变形的花园……水花园"，"公路花园……为空间，时间和运动而设计"，"花园城市"。40年后的2008年，全部原始设计稿与文字首次在美国完整出版——见吴欣著《重塑现代性：帕特里夏·约翰逊的〈家园与花园〉设计方案（Reconstruction of Modernity: Patricia Johanson's House & Garden Commission, 2 vols）》（上下册）

兴趣,并开始探索一种将自然与城市环境连接起来的混合艺术时,已经是位得到认可的极简主义艺术家。能简短地描述一下您本人思维的转变过程么?

约翰逊:1969 年,那份委托完全出乎我的意料。当时我是个职业画家,对景观设计几乎一无所知,只能边做边学。当自学成才的人被迫去做一些事情,比如设计一个花园,有趣的是你没有背景、没有可以遵循的模板、眼前没有任何例子。这迫使你要有非常的创造力并且亲力亲为。我受过艺术方面的严格训练,所以我懂色彩、构图,我也懂雕塑。因为喜欢建筑和工程,所以我也知道一些结构方面的知识。但在碰上景观设计的时候,所有事物看起来都是完全不同的东西。其实在今天看来,我觉得艺术和景观设计并没有那么大的差异。

想当年,年轻的我一开始真的是不知道如何是好,所以就尝试了各种方法。我广泛阅读了关于花园、庄稼、树、蕨类、苔藓类、地衣类等几乎所有关于可生长物体的书。我不担心构图和设计之类的事情。当然,我开始与阳光和水为伍,因为需要培育植物。这就引发了水和生物群落的问题,然后自然而然地过渡到了各种环境问题。我没有遵循当时景观设计师的范例,做出规范的景观设计,而是转而探讨可持续性、生态社区、空气和水污染。20 世纪 60 年代的热门话题是瑞秋·卡森(Rachel Carson)和简·雅各布斯(Jane Jacobs)的著作,和所有关于环境污染、城市衰退的讨论。这些正是我当时所做的基础阅读。所以我并没有像一个极简主义艺术家或者一个受过训练的景观设计师那样工作;而更像是一个对环境问题感兴趣的业余爱好者。这就是我的起点。举个例子来说,《家园与花园》的提案中有许多是关于湿地水处理的。1969 年的美国,当时并没有任何湿地污水处理厂,我只能是从理论或者说推测的角度来看待它。但我从未怀疑过它的可行性和科学性。我坚信它是可以实现的。正如我们今天所见,湿地方案在国际上已经被视为废水处理的常规选项之一。

吴欣:我可以说您与景观的缘分是从关注环境问题开始,而非从形式开始的么?

约翰逊:我想你可以那么说。直到今天,无论什么项目,我总是在项目伊始花大量的时间去研究与其相关的环境和生态。

吴欣：然而，许多人对您的设计作品印象最为深刻的是它出人意料的形式，利用自然形式的想法是如何成为您艺术标志的？

约翰逊：重申一下，我不像大部分景观设计师那样受过正规训练。艺术家的训练引导我观察周围世界的一切。图画是我真正的语言。我描绘所有的事物。我从一片叶子开始画起。当你想画一片叶子的时候，你就会注意到叶子上的所有细节——叶柄、主叶脉、叶脉、叶片……当你尝试画叶脉的时候，你就会想到一个给水系统，会想象叶子如何把水输送到需要的地方。如此，你可以开始绘出图案，以及图案的各个层面。我自此开始，作为一个艺术家简单地描绘着自然界所呈现出的事物，同时思考着人类社会中的许多问题。当然有时也会回到环境问题上。在自然界，所有图案都是美丽实用的。例如，当我环顾房屋四周的时候，在那些形状漂亮的叶子上，可以看到昆虫留下的足迹……你可以看到，我与自己的对话包含着自然的事物和现象，而这总是将我的注意力引回到环境问题上。

吴欣：这听起来很像新儒家看世界的方式——"格物致知"。

约翰逊：千真万确。无论是在艺术方面还是在伦理方面，大自然都有很多值得学习的地方。

自然形态并不是唯美的，而是生动有意义的；大千世界中存在着生命的秩序和逻辑。我把植物的结构和组织转化为艺术，并以这种方式努力探索有意义的形式，这说明了为什么我的一些早期景观设计是基于植物形状的。我曾经设计了一个单细胞衣藻形状的公园，因为这种衣藻给了我灵感，它是一种社会的存在形式，一个复杂而高度组织化的微型世界，就像我在儿时就很喜欢的奥姆斯特德（Olmsted）设计的纽约中央公园，蝴蝶景观也是如此。在动物王国，一个物种的身体形态是进化的结果，通常也是一种适应其环境的伪装方式和生存策略。所以我们可以理解，从这个角度来说，蝴蝶身体上的图案形态本身就是一种景观。

吴欣：您在许多讲演和文章中强调，您的艺术不是关于设计到的而是关于没有被设计到的地方。您能详细解释一下这个声明么？

约翰逊：我的景观中最重要的部分和它们成功的关键在于我没有设计的部分。现在我设计项目规模很大，也在地面上留下很大的图案，但是我不打算把它们做成用来观赏的图案。许多人想要看到这些图案，所以他们乘直升机在空中拍照，并相信那才是景观设计中应该被看到的东西。情况并非如此。相较而言，我更注重在体验它的实质和看到它的形象之间建立联系，以及作为人类你要如何通过探索自然发现这一点。我想要人们去探索，而不只是看。

当你行走在我的某座雕刻作品上时，你正以一个穿越空间的舞者的方式画出一个图案，同时获得着不同层次的感知。例如，当人们遥望达拉斯雕刻的时候，它看起来很大，很壮观，它是个艺术品；当人们朝它走去，到达雕刻上，跑在上面，他们就看不到这个艺术品了。随后他们的注意力就转移到周围自然中的小东西上了。一只从水中冒出来的蟾蜍，一只蜻蜓，或者一朵花，不管是什么他们都会感兴趣。不错，确实有一个设计好的图案，这你可以在项目图纸上看到，但是在现场，它如此之大以至于你无法窥得全貌。此外，很可能你最后根本不会介意这一点，你的注意力都转移到其他对你重要的事情上了，而不是这件艺术品。你会转而去注意那些艺术品周围的野生动物群和大自然。

吴欣：那您如何总结您的设计方法呢？

约翰逊：对于每个设计，我都有或多或少相同的议程，或多或少相同的目标。就是说，设计意在美学体验，然而它必须首先满足一些功能。我所做的项目通常涉及一些与公路、污水处理厂或者采矿项目相关的超大型功能性基础设施。所以，总是要有一个主体功能部分，然而，我也另外引入一些功能性相对较弱的部分，因为我想要得到于人有益的景观。比如，我大多数的项目都融入了公园和步道。而且我也希望这些项目成为野生动物的聚集地。所以我尽量设计生活型的生态景观，在那里动物和植物可以相生相克。所以实际上它更多的是在重建一条食物链，而不只是景观设计。这就是我的设计方式。

所以尽管存在所有这些层次，即使每个项目可能都是不同的，但它们也会看起来像一件艺术品，因为任何项目都必然有其形式。我首先尽可能地做好一个生态学家、一个科学家,具体问题具体对待,比如,知道什么植物能把特定污染物从流入场地中的水中清理掉。

吴欣: 这样说来您其实是要淡化形式在设计中的支配地位!这很有意思,因为似乎很多人最关注的就是您设计中的形式。他们认为您的方法是形式主义的,因为整个项目的总体图案看起来是靠您所使用的形式来支撑的。

约翰逊: 我完全理解为什么人们会这样认为,但是他们错了。我给你举个例子。当我受邀为达拉斯美丽公园潟湖做设计时,我脑子里并没有它的形式,在设计过程中我很晚才会考虑形式问题。基本上,我做了许多植物和动物方面的研究,因为当我到那里的时候,潟湖里已经没有任何动植物了,所以第一个想法就是把它们找回来。我研究了德克萨斯州、达拉斯本地有哪些动植物符合要求。我还去了达拉斯自然历史博物馆(就在潟湖旁边),了解潟湖可以提供哪些有良好教育性的活展品给策展者,对玻璃箱子里的展品进行补充。我说几个问题,借以强调在我的解决方法中功能问题的重要性。首先,堤岸侵蚀问题。湖岸线每年正以惊人的速度被侵蚀,湖岸缩小,寸草不生。整个岸线变得光秃和贫瘠,然而我们需要岸线上的植物来支撑微生物和动物的生存。其次,洪水问题。湖岸线易受洪水威胁,所以,我意识到潟湖可以在降暴雨时储水,之后将水排放到小溪中。当然还有水质问题。岸边有草,还有公园部门给草施肥。下雨的时候,雨水会将肥料冲到潟湖中,导致藻类的爆发。所以另一个目标是提高水质。

我从未在设计前数月考虑过某图案。实际上,一旦我了解到问题就会开始研究方案。为了防止湖岸受侵蚀,就要削弱波浪的侵蚀作用,所以就需要在受到侵蚀的整个岸线上构造一个连续的防线。然后,再一次提及,为了给动物创造微型的栖息地,需要在广阔的潟湖上营造小空间,因为比如像龟之类的动物需要藏身的地方。它们喜欢生态龛,喜欢食用特定的植物,所以就开始处理水生植物的问题。考虑到建立食物链,就需要不同的水层,需要一些浅水,一些大约半英尺(约15cm)深的水和一些非常深的开放水域。我就是这样开始设计一个生态景观,来吸引那些可以相互影响的物种。然而,这里仍旧没涉及图案。

在美丽公园，潟湖边沿的长度是另一个问题，因为潟湖有 5 个街区那么长，而且人们无法穿越。所以我的另一个目标是在水面上建些桥和路。在方案满足了所有这些意图之后，我才坐下来考虑可以用什么形式。因为我研究了如此之多的德克萨斯州植物，所以就选择了其中两种。首先我选择了一种蕨类植物（凤尾蕨，*Pteris multifida*），它有许多不同的小叶，因此基本上你可以把它们扭起来，放大，把它们做成路或岛。所以，如果洪水和干旱时水有所涨落，通道就可能是不同的，有时人们是踩着石头过去的，而有时则有一条完整的路。其次选择了一种本地水生植物（扁叶慈姑，*Sagittaria platyphylla*），它有着盘曲的根，可以用来削弱水浪的作用。因此，我最终采用的图案是与地点、当地生态，以及具体功能性问题相关的。我不会在其他地方再用这些图案。简而言之，当设计一个项目的时候，我首先分析功能性问题，之后选择在当地意义重大的图案，最后把图案展开，并把设计的各种目标整合并完成一个统一的形式。

吴欣：非常感谢您的解释。这有助于我们理解在您的设计方法中解决问题是优于创造形式的。那么，您可否谈谈您在韩国公共公园项目中所使用的龙形？这个图形不是用来回顾历史或者当地文化而非生态的案例么？

约翰逊：同样，它更类似于前面的例子，而不是像最初看起来的那样。我对龙、天龙和水龙的整个构思都十分着迷。大规模的自然现象有许多不同的表现。当我开始探究这一点的时候，越来越觉得龙并不是一个神话而是某种真实的存在。我相信龙真的存在。关于龙的整个故事是想告诉我们有关环境的一些重要事情。这就是为什么我选择这个图案，至少算是一部分原因吧。

我还选择了这个概念：龙从未现其全貌，而只有一部分，如云龙的龙爪，或者龙的其他部位。我发现这也很诗化，对于具体的设计我不知道要说多少，但在看起来这个特定空间里，能量的流动非常重要，而龙则是其中的关键之所在。这也是为什么在龙脊的顶端——我们称之为"龙径"——设置了一个填满水的池塘作为龙眼。当太阳映入池水中的时候，天龙和水龙就合二为一。因此生态学中试图在概念上实现的：将自然的不同部分结合到一起，将以一种非常真实的方式实现。所以我是根据当地的实际情况,而非将其作为一种神话去使用这个图案的。而且我们可以看到在中国和其他地方的考古学家越来越多地发现禽龙的骨头。如果你想到禽龙,一种飞在天上的爬行动物,你就会觉得龙不只是个神话,而更多的是先祖记忆的一部分:一种形式,无关大小。

吴欣:我明白了:一个民俗学者可以借用当地文化中独有的神话、观念或者图案,将其作为一种形式强加于景观之上,妄图重现过去,但您的态度是完全不同的,而且最终看来,您对当地文化更为尊重。您寻找当地文化中的观念或者故事,将之作为现今生态学和环境著述中提出问题的一种答案,之后您向"自然巨著"借用生动形态的轮廓,而这个形态既涉及了当地的传统文化,又涉及了当今的环境著述。在其他方面您的设计态度也与形式主义和民俗主义泾渭分明,因为您的设计态度中首要的是关注某种文化变迁,这种变迁有助于建立一种新的理念,尊重众人对大自然的兴趣。所以不管是在西方国家或者是非西方国家,您都不会作出相同的设计,但您会保持相同的目标和很大程度上相同的手法。设计与传统和乡土的关系目前在中国很受关注。我知道您曾受邀来中国演讲并接受了电视台的采访,您认为中国设计师应该如何提炼他们自己的文化遗产,进而设计出好的项目呢?

约翰逊:但愿我可以!不过我想说说为什么很难给中国设计师建议。对于任何委托,我作为一个设计师(与当地人相比)总是局外人。作为一个局外人,我知道我有很多要学。所以,我总是试着找到熟谙当地景观生态的人,去了解当地人怎么想,想要什么。因为每个地方都是不一样的,而且文化差异会很大。即使在美国本土,我在加州做设计的时候,我并不是一个加州人;而当我在盐湖城做设计的时候,我也不是来自盐湖城的。所以我力图花尽可能多的时间去了解这些地方,了解那里的人们是怎么想的,他们想要什么。对我来说设计一直是一个学习的过程。

每个设计师都知道,当你提出一个设计提案的时候,要忠于你的想法,即使由于实际情况和政治原因,这并不总是那么容易。我还认为开发一个满足实际需要的合理项目是最重要的,不仅要考虑功能和人的问题,而是将这个充满生命的世界作为一个整体。与人类中心说相反，我认为我们做设计不该只为人类，也要利于野生动物的生存。这样做，将对人类福利起到超出我们想象的帮助作用；如果野生动物消失了，我们的生活将极度贫乏。请把这一点加入到你对我设计方法的总结中。

吴欣：为野生生物做设计。那么，您并不满足于保护，而是想通过设计同时提高人类和野生动物的生存条件。这呼吁人们关注野生自然物种的生存和发展问题，然而，这又会引起另一个问题，因为当地人对于自然可能有很多意见分歧。设计中碰到这些困难时，您如何处理？

约翰逊：是的，太对了。关于自然和野生生物，在任何当地社区都存在意见分歧。所以我很小心地避免一些代表既得利益群体的意见。一些人或者团体可能想要只满足他们自己利益的设计，而不是本质上的公众利益。我倾向于不去听他们的。我宁愿尽量包容，为所有团体做设计。比如，在美国，有很多地方团体属于一个鸟类保护学会——奥杜邦学会（the Audubon Society）。这些几乎都是有钱人，他们只关心鸟。他们经常为了保护鸟而将其他所有野生生物弃之不顾。这是非常危险的，因为他们盲目地更改自然物种平衡。相反，我却支持多样性：生物多样性和文化多样性。

19世纪，一个非常著名的景观设计师弗雷德里克·劳·奥姆斯特德（Frederick Law Olmsted），他的作品遍及美国，创造了一系列伟大的城市公园。从建成一直到现在这些公园人人受用，人们爱它们。这是那种有重大意义的设计项目：既满足所有当地人口目前需要，又能够延续至未来的项目。我们也应该为我们的后人做设计。

吴欣：我也和您一样希望有更多的设计师像奥姆斯特德那样为"后人做设计"。让我们回到访谈一开始的论题。现今在景观设计领域有越来越多的设计师介入"绿色基础设施"开发项目——绿化是个很时髦的词——比如模仿平原上的河流，在城市中设计一个河谷，以实现适用于这种自然环境的可持续性。您的设计不模仿自然形式,而是引入巨大的象征性形式,往往将放大了的场地上现有物种的轮廓融入设计中。您常常将设计项目比喻成活的孩子,能以佩塔卢马湿地水厂公园为例,解释一下您的设计艺术作品如何能自我持续和发展么?

约翰逊: 河水流动不是一个技术性过程,而水处理却是。在任何环境下,都有自然过程随着时间流逝塑造着河流;对于水处理则不然。但是城市需要清洁居民和工业排放用水,以及城市的雨水,那么这种水处理就要有自己的形式。在佩塔卢马,设计是按照当地的濒危物种,一种小型盐沼鼠的形状做的。这个是我们为该设计提出的第三个图案,因为在研究过程中,一些状况导致了"处理流程"也就是处理污水的技术过程和自然过程发生了变动。为了实现可持续性,首先我们尽可能降低能耗。所以整个工厂的水流由重力产生。这只是我们送水时的一个例子,否则水将一直朝下坡流。其次我们考虑到要节约建筑材料。那里本应使用更多硬件和技术设备来处理污水的。再次,工厂建得离地面很近。我倾向于选择真正简单、基础的东西,所以我设计了土池。它们非常易于建造,成本低且易于修理。这样一个项目,在它整体环境方面有更多的自我可持续性,因为加州的这片区域遭受过数次大地震,一旦溃坝就会肆虐下游。土池有很好的抗震性,而且这些池塘的设计保证即使最坏的事情发生,也就是一旦有地震,这些水——已经足够清洁,可用于农业或者排放到自然界——就会流到几百米外的河中,流下去并弱化在河里。所以基本上是处理流程的技术过程与环境和景观设计融合在一起保证了这个项目的自我可持续性,而不是它们所呈现的形式。

吴欣: 您已经谈到了这个项目的技术和环境方面;它的艺术性呢?

约翰逊: 艺术不是一个附加的装饰;它与设计是一体的。让我举个例子。这个项目有一部分用于处理当地公路和邻近商业中心过来的雨水。这是含有油污和化肥的污水,不允许直接流到佩塔卢马河里。我设计了一个独立的过滤过程,驱使水流经一系列覆有植被的小池塘,整体形状是一朵牵牛花。这样就形成了一个有着颜色各异的多种水生植物的格子,与中间的水池共同构成了一朵牵牛花的图案。现在美学部分当然就是颜色和图案了;功能部分是对过滤植物的选择,每种选定的植物处理水中特定的污染物。所以，如果你在邻近池塘护堤上看这个花的图案，你就同时看到了艺术图案、颜色、植物和技术过程。同样的道理可以被用来解释盐沼鼠形深度处理池的魅力——环境艺术，色彩美学和净水功能三位一体。

吴欣：您期待游客如何去欣赏佩塔卢马的项目？他们应该对水处理流程的设计有所理解么？或者是将注意力转移到对盐沼鼠的保护上？又或者您有其他期待？

约翰逊：仁者见仁，智者见智吧。如果你对野生生物感兴趣，你很可能来这看它们；如果你是个徒步旅行者，对公园和小路感兴趣，你可以跳上自行车或者沿着小路玩耍；如果你是一个植物学家，那里有许多人们喜欢的美妙植物。我真的不确定人们初次会怎样看待它。很难预计。

目前并不是所有的艺术都完工了；我们还将在完成的池子里撒入彩色的沙子。这个完成后，你就可以看到水中映射出的色彩。这非常重要，因为它会将人们的注意力吸引到水流上，以及水不断变化的外观上。这也是一个强烈的视觉现象，而你也许会期望这将引向另一次审美，而不是技术性的水处理池。我上次访问的时候，看到有几个画家在一个水池岸边摆放了画架，正投入地写生。克劳德·莫奈（Claude Monet）曾在他的花园里种了一株睡莲，为的是自己画出日式环境下的睡莲；佩塔卢马的一些居民将这些池塘中的岛视作激发他们对大自然的认识和见解的地方。这些岛是工程的一部分，因为它们将水流由上一个完成处理的池塘导向下一个。所以它们是技术基础设施的一部分，但它们也被设计用于吸引野生动物，为鸟类提供筑巢区。而且，我想在道德层次上，这是很重要的，每次我们建造什么东西的时候，想想我们侵占了野生动物的土地，我们在道德上有责任为给野生生物带来的不便做出弥补。这是我选择植物背后的动机之一。我选择它们为所有鸟类和动物创造新的栖息地和筑巢区。这些动物不是俘虏；它们来是因为受到植被的吸引，受到底泥的吸引，受到它们猎物的吸引。动物可以为人们提供娱乐，人们可以让孩子们看水獭，或者狐狸，让他们了解院子以外的动物。越是自省就越会超越这个层次。他们会看到这个项目在为野生生物设计时，并不是从你看到的开始，而是下到了微生物水平。微型动植物更为微小的生态系统,实际上负责了大量水质问题和处理。它们将通过个人的方式企及一个新的道德水平。

吴欣: 您将欣赏景观视作一个随着时间发展的过程:景观设计应该给予快乐,并引起对野生生物的审美和激发对环境问题的道德参与。这是否也意味着一种新的设计实践呢?

约翰逊: 我想景观设计师可以有很大的贡献,因为他们与城市环境建设的关系如此紧密。我强调一下时间的作用。一旦某种类型的景观开始形成,我们会训练眼睛去喜欢它。修剪景观是所有景观设计师都用在住房和公共建筑周围的一种类型。美国的自然景观与修剪景观之间的矛盾之一,就是人们想看到草坪,他们想看到修剪的草,这样最终就会用到杀虫剂,而杀虫剂会流到水中导致所有后续的问题。当面对自然景观时,人们会觉得那只是一片旷野,看起来像杂草,似乎无人照料。然而随着岁月的流逝,也许正在改变。所以,景观设计师有责任向可持续发展的方向训练人们的视觉。要做到这一点,他们不该只为了愉悦眼睛,而过多地从视觉方面去考虑种植。当今世界,人口膨胀,所有资源都变得紧张。我想民众们应该参与到多功能、多用途的公共基础设施的开发中,这种基础设施为野生生物的发展提供了一种新的重要支持,同时也为人类提供了一种高效的自我可持续服务。

这意味着在任何设计项目中都要大力鼓励广泛的公众参与。即使这很困难,要花费很多时间,即使没人能百分之百得到他想要的,我认为相比于设计师单独工作,相关参与者共同努力将成就一个更好的项目。

就我个人而言,我喜欢做新的有功能性的项目。每个项目成为某一特定条件下的范例,就像我现在正为宾夕法尼亚州一个有毒的煤矿场所做的复活改造项目。这是一个你什么都种不了的地方,水退到地表以下。我想把那片土地找回来,让它重新变得丰饶、美丽和实用。我想要人们能够去那里,看水、在河里游泳,而他们现在却无法这样。这些项目正是我想做的类型,即使我做的不多。每次我选择一个项目,我都尽量去探索那些真正有重要性的论题。只要可能,我就会继续坚持下去。 (李云圣 译,吴欣 校)

Patricia JOHANSON
—between public art and environment infrastructure

Patricia Johanson is an American environment artist and landscape designer, and one of the pioneers of what has come to be called ecological design. In her long career of 50 years, she has been insisting on the environmental and social function of landscape art and architecture, in particular their function in dealing with problems of urbanism and reforming aesthetics of nature. She started her career as one of the early minimalist painters in post-WWII New York. In 1969, the course of her art was dramatically deflected by an unexpected garden commission from the *House & Garden* magazine. It resulted in 150 garden proposals and seven essays of new garden genres. The 1969 *House & Garden* proposals explored many issues that have become influential nowadays in contemporary landscape architectural projects, such as the green roofs, the highline and ecological infrastructure, etc. Although unrealized, the commission opened new doors for the artist and pointed to the possibility towards what she has always strived for—"Framing the World as a Work of Art". Since then she turned exclusively to landscape, creating a highly personal hybrid art that incorporates art, landscape architecture, civil engineering, environment and ecology. Her work has drawn more and more attention since the 1980s. Recently, she worked intensively with civil engineers in large scale infrastructures such as waste-water treatment plants, wetland parks and trail systems. This interview start from the *House & Garden* proposals, but will focus on her recent built projects.

Johanson's career has revealed an unknown development of the post-WWII American Art, putting forth a renewal of landscape art in defiance of well-known cultural trends: formalism and modernism, earthworks and environmentalism. Lastly, Johanson is a lover of Chinese art and philosophy. She translates non-Western (in particular Chinese and American Indian) natural aesthetics and ethics into answers to contemporary urban issues. Like a good traditional herb doctor, she believes in healing the earth through reenacting the internal vitality of the place, the people and the culture. To reconnect the fragmented mountains and waters is the core of her belief.

Xin WU(WU hereafter): Ms. Johanson, I hope more people will learn from your new project in Petaluma[①]: it is a wonderful example of an infrastructural waste water treatment plant turned into a recreational wetland park.

Patricia JOHANSON (JOHANSON hereafter): In many ways, this new facility introduces a common problem and an unusual solution that might resonate throughout the United States. Petaluma is a small city in Sonoma County, California, with about 50,000 inhabitants. Like many other cities in this country, it had an old water system that was working beyond its capacity. The city council was concerned about the scarcity of drinking water, the environmental quality, and the endangered species; meanwhile it wanted the administration and citizens to reach a greater level of awareness of these issues. I was first invited to visit in August 1998, and then commissioned to work with Carollo Engineering at the design of a new wastewater treatment facility. It has just been completed and opened in July 2009.

WU: This is a wonderful example of a functional landscape. In China, more wetlands are created now, but little attention has been given to their role in water-purification. But before we enter the details of this new project, I would like to first locate it in a historical perspective. You were already an established minimalist painter in the mid-1960s in New York, when you became interested in landscape through the

---

[①] This new facility overlays public landscape art, infrastructural sewage treatment, and habitat restorations, and urban agriculture. Embedding major urban infrastructure within living nature, Johanson created a multi-purpose public landscape, using constructed and natural wetlands to provide three miles of public trails for recreational use, educational programs, nature study and tourism. The park coincides with the $150 million dollar Ellis Creek Water Recycling Facility that simultaneously processes human sewage and stormwater runoff, creates wildlife habitats, serving as a highly visible model for converting wastewater into drinkable water.

House & Garden commission [1], and started to explore an hybrid art to connect nature to urban environment. Can you describe briefly the development of your ideas?

JOHANSON: Well, the commission came to me totally unexpected in 1969. I knew little about landscape design at the time and I had to learn everything en route. One of the interesting things for somebody self-educated when pressed to do something, like design a garden, is that you do not have background, you do not have any patterns to follow, and you do not have any example in front your eyes. It forces you to be very inventive and to figure things out on your own. I was trained as an artist, so I knew about color, composition, I also knew about sculpture, I knew a little bit about structure since I liked architecture and engineering. But when it came to working on landscape, it seemed to be something completely different. Now I actually think art and landscape design are not so different.

Since I had really nowhere to turn, I turned every-where. I kept reading broadly books about gardens, plants, trees, ferns, mosses, lichens... just about anything, anything that grows. I was not worried about the composition and design, or anything like that. Of course, I became involved in sunlight and water, because you need to feed the plants. This led to water issues and fauna issues...and automatically to all sorts of environment issues that were very important to the 1960s and post-WWII Americans. So instead of following the example of landscape architects at that time, and produce formal designs of landscape, I moved towards sustainability, ecological community, and confronting the poisoning of air and water. The hot topics in the 1960s were the writings by Rachel Carson and Jane Jacobs, and all the discussion

---

[1] As the result of the *House & Garden* commission 1969, Johanson produced 150 design proposals under seven new garden genres: "Gardens by the Mile...the Line Garden"; "Gardens That Are Out of Sight...the Vanishing-Point Garden"; "Nature and Anti-Nature...the Artificial Garden"; "Gardens of the Senses and the Mind...Illusory Gardens"; "Gardens that Nourish, Drift, and Transform...the Water Garden"; "Gardens for Highways...Designing for Space, Time,.and Motion"; "Garden-Cities". These 1969 drawings and texts have been published in a 2-volume monograph in English—Xin WU, Reconstruction of Modernity: Patricia Johanson's *House & Garden* Commission, (Washington DC: Dumbarton Oaks; distributed by Harvard University Press, 2008)

about pollution, urban decline. These were basically the reading I was doing. So I was not working as a minimalist artist or a trained designer; I worked more like an amateur person who was interested in environmental issues. This is where I come from. For example, many of the *House & Garden* proposals were about wetland water treatment. There wasn't any wetland sewage plant built in the United States in 1969, I was looking at it from a theoretical or speculative perspective. I have never had any doubt that this could be done. But I was able to propose because I was not constrained by what could or could not be done. I firmly believed it could be done. As we can see now, wetlands have been routinely considered as one of the options for waste water treatment.

WU: Should I say that you started from research on environmental issues instead of form?

JOHANSON: I think you can say that. To this day, I always spend a lot of time researching the environment for any project.

WU: Yet many people are most impressed by the dramatic forms of your work. How did the idea of using natural forms come to become the signature of your art?

JOHANSON: Again, I did not have the kind of training most designer had. My own training leads me to draw everything. That is my true language. So I began to draw all of these things. I began to draw a leaf. When you think about drawing a leaf, it leads you to all little things in a leaf—petiole, mid ribs, vein, lamina… When you try to draw the leaf veins, it leads you to think of a water distribution system, and how the leaf transports water to where it is needed. So you begin to develop pattern, and layers in the patterns. I started out, as an artist to draw simply something present in nature, while thinking of a lot of issues in human society. And of course, it also went back to environmental issues. In nature, there is no pattern, which is not beautiful and functional. For example, when I look around my house, the forms of leaves are pretty; you might see trails on them left by insects. Then you may wonder what is feeding the leaf and what are the relationships between the insect and the leaf…You can see how, in my dialogue with me about natural things and phenomena, it always led back to my environmental concerns.

WU: That sounds very much like the neo-Confucian way of seeing the world—"ge-wu zhi-zhi; investigating things [and] extending knowledge".

JOHANSON: That is absolutely true. There is much to learn from nature, whether artistic or ethical. The forms are not aesthetical but living and meaningful; there is the order and logic of life in a myriad world. Some of my earliest landscape designs based on plant forms resulted from efforts to explore 'meaningful' form by translating the structure and organization of plants into art. I once designed a park after the form of single-celled lichen Chlamydomonas, because it inspired me as a social form of existence, a complex yet highly-organized miniature world, just like the New York Central Park by Olmsted which I have admired since my childhood. This is the same for the butterfly landscapes. In the animal kingdom, the body patterns of a species resulted from evolution, often a way of camouflage adapting to its environment, which is a strategy of survival. So, we can understand that literally the form of the butterfly IS the landscape.

WU: You have written and stressed in many lectures that your art is not about design but what is not designed, can you explain this?

JOHANSON: The most important aspects of my landscapes and the key to their success lie in the parts I do not design. I design very large projects now forming very large figures on the ground, but I do not intend to offer them as an image to be looked at and appreciated. Many people want to see the image, so they fly over in a helicopter to take a photo and believe this is what is to be seen in this landscape design. Such is not the case. I am much more involved in the interplay between experience of what it is and the image of what it is, and how you find that out as a human being through exploring in nature. I want people to explore, more than to just see.

When you walk on one of my sculptures, you exactly walk a pattern in the way a dancer moves through space, achieving different levels of perception. When people see the Dallas sculpture from a distance, for example, it looks big, grand, it is art. When people reach towards it, get onto the sculpture, run onto it, they cannot see the art any more. Then their attention is drawn to small things in nature around them. This could be whatever interests them—a toad that comes out of the water, a dragonfly, or a flower. It is quite true that there is a designed image, and you can see it on the drawing for the project, but on the site it is so large that you won't see it in full. Furthermore, probably in the end you will not mind, your attention is drawn to something else that is important to you, instead of the artwork. You will turn to the flow of wildlife and nature around the artwork.

WU: Well, then how would you summarize your design method?

JOHANSON: For every design I have more or less the same agenda, more or less the same intent. And that is, it aims at aesthetic experience, but then it must first and foremost fulfill some function. Projects on which I work usually concern very large functional infrastructure connected to highways, sewage treatment plants, or mining projects. So, there is always a primary functional component, but then I also introduce another lesser functional component, because I want the resulting landscape to be useful for people. So for example, most of my projects have parks and trails woven into them. Besides, I also want them to be sources of wildlife. So I try to design living kind of ecological landscape where plants and animals are basically consuming each other. So it becomes more of a food chain actually than a landscape design. That is the way I use to design.

So there are all these layers but each project even though it may be different, also looks like a work of art, since any project must have form. But I mostly try to be a good ecologist, to be a good scientist be able to target specific solutions to specific problems, for instance, to know what plants will clean the water from the particular pollutants found in the water running into the site.

WU: So you downplay the importance of form! That is very interesting, since many people seem to pay mostly attention to form in your design, considering that your approach is formalist since the form you used appears so important in holding together the master plan image of the whole project.

JOHANSON: I totally understand why people would think that, but they are mistaken. I will give you an example. When I was invited to build a sculpture for Fairpark lagoon in Dallas, I had no form in mind, and I only turned to issues of form very late in the design process. Basically I did a lot of research about plants and animals, because there was no longer any plant or animal life in the lagoon when I came there, and my first idea was to bring them back. I researched what water plants and animals native to Texas, to Dallas, feed upon. I also went to the Dallas Museum of Natural History (just alongside the lagoon) to know what good educational living exhibits the lagoon could provide the curators with, to complement what they had in glass cases. Let me name a few issues to underline the importance of functional problem solving in my approach. There was bank erosion. The shoreline was eroding at an alarming rate every year, and the shores were compacted so nothing could grow there. The whole shoreline was bare and sterile, and yet you need plant life along the shore to support micro organism and animal life. Then, there were flooding issues. This was an area that was prone to flooding, so I realized that the lagoon could store water during a storm and release it afterwards into the creek. And of course, there was also a problem of water quality. There was grass on the shore and the park department was fertilizing the grass. When it rained the water would wash out the fertilizer into the lagoon resulting in a big algae invasion. So another goal was to improve water quality.

And never once in the first many months of designing did I even think about an image. Actually once I knew what the problems were I started working on solutions. Well, in order to prevent shoreline erosion, you want to break up the wave action, so you want something that would form a continuous line of defense across the shoreline where the waves are breaking. And then, again, to develop micro-habitats for the animals, you want to introduce smaller spaces in the large expanse of the lagoon, because animals like turtles, for instance, like hiding spaces. They like niches, they like certain plants to eat, so, you start dealing with water plants. Since you are concerned with building up food chains, you want different gaps of water; you want some shallow water, some water that may be a half a foot deep and also some very deep open water. And this is how you begin an ecological landscape were you can attract many different species, that would interact with each other. Yet, still there was no image.

Distance around the lagoon was another issue at Fair Park since it is five blocks long and people could not cut across. So another goal of mine was to create bridges or paths over the water. After all these intentions were developed, I sat down and pondered what form I could use to accomplish them. And because I had been studying so many Texas plants, I chose to use two of them. First I chose a fern (*Pteris multifida*) which has many different leaflets, so basically you could twist them around and blow them up to make them into many different paths or islands. So if the water rises in a flood or falls in a drought, you prepare different access routes, and sometimes people just go on stepping stones, and sometimes there is the continuity of a full path. Second, a native emergent plant (*Sagittaria platyphylla*), with all its twisted roots, could be used to break up the wave action. So the images you put in are related to the place and its ecology, as well as to the specific functional problems. I will never use them anywhere else. In short, when designing a project, first I analyze the functional problems, then I pick an image significant within the local area, and finally I deploy that image to accomplish and bring together into a unifying form the various purposes of the design.

WU: Thank you very much for this explanation. It certainly helps understand how problem solving takes precedence over creating form in your approach. So, to pursue this clarification, could you say a few words about the dragon form you used in your project for a public park in Korea. Isn't it a case when the image harks back to history or local culture rather than ecology?

JOHANSON: Again, it may be more similar to the preceding example than it seems at first glance. I was fascinated with the whole idea of the dragon, the celestial dragon, and the water dragon. There were different manifestations of large scale natural phenomena. As I began to explore that, it seemed more and more that the dragon was not a myth; that it was about something real. I was led to believe that the dragon actually existed and that the entire story about the dragon was really telling us something important about the environment. That is, at least in part, why I selected this image.

And the idea that the dragon never manifests its whole image but only a part, as the dragon claw of the cloud dragon, or some other aspects of the dragon. I found that also very poetic and I do not know how much I want to talk about this particular design, but it seems that this flow of energy, was so important in this particular space, and that the dragon was the key to it. That is why at the very top of the spine—let us calls it the "dragon trail"—I placed the dragon's eye, as a pool filled with water. The sun, coming down into the pool of water, would unite the celestial dragon and the water dragon. Thus it would achieve in a very real way, what ecology tries to achieve conceptually: bringing together the different parts of nature. So I did not use it that image as a myth way, but rather as local reality. And more and more, we see in China and elsewhere archeologists finding bones of therosaurus. If you think about a therosaurus, which is a flying reptile, it makes the dragon much less a myth than a part of ancestral memory: a form, irrespective of size.

WU: I see: a folklorist would borrow any myth, idea or figure particular to local culture to impose a form upon the landscape in the vain hope of reviving the past, but your attitude is completely different, and finally more respectful of local culture. You look for ideas or narratives belonging to local culture that express an answer to problems addressed at present by ecology and environmental discourse, and then you borrow from the "great book of nature" the outline of a living shape that speaks both to traditional local culture and to the contemporary environmental discourse. In other terms your design attitude stays clear of both formalism and folklorism, because it is first and foremost concerned with cultural change that will contribute to a new engineering respectful of all people's interest in nature. So, you would not produce the same designs in a Western or a Non-Western country, but you would maintain the same goals and, to a large extent, the same method. This is of great interest in China at present. Since I know, you were invited to give talks and were interviewed on TV in China; I would like you to explain in a few more words how do you think that Chinese designers should hone their own cultural heritage to design better projects?

JOHANSON: I wish I could! Let me say why it is difficult for me to advice Chinese designers. In any commission, I always come in as an outsider. Thus, as an outsider, I realize I have much to learn. So, first I try to find local people, who know a great deal about the local landscape, housing design, how local people think, what they want. It is different in every place, and cultural differences can be huge. Even in my own country, when I design in California, I am not a Californian; and when I design in Salt Lake City, I am not from Salt Lake City. So I try to spend as much time as I can learning about those places, and trying to understand how people think there and what they want. Design is always a learning experience for me. So, I cannot give any specific advice to Chinese designers.

As any designer already knows, you need to be true to what you think when proposing a design, even though this is not always so easy for practical and political reasons. I also think that it is most important to develop a sound project that fulfills real needs taking into account not only functional and human issues, but also the living world as a whole. I would suggest, against anthropocentrism, that we should not just design for human beings, but also in favor of wildlife. It would be a great impoverishment of our lives, if wildlife were to disappear; and moreover it may contribute more than we can imagine to human welfare. This is the point I would like to add to your summary of my approach.

WU: Designing for wildlife. So, you are not satisfied with conservation, and you want design to improve the living conditions of both humans and wildlife. This call for attention to the issues of survival and development of natural species of wildlife, however, raises another issue since local people may entertain conflicting views about nature. How do you deal with these dilemmas in your own design?

JOHANSON: Yes, it is very true that there are conflicting views about nature and wildlife in any local community, so I am careful to avoid views that represent some vested interest. Some people or group may want design to accommodate only their own interest, not the general public interest in nature. I tend not to listen to them. I rather try to be very inclusive and to design for all groups together. So for example, in America, there are many local groups, belonging to a bird protection society, the Audubon Society. These are mostly very wealthy people, who only care about birds. Often, for the sake of protecting birds, they would cast aside all other wildlife. This is very dangerous because it blindly aims at modifying the balance of natural species. To the contrary, I really support diversity: bio diversity and cultural diversity.

In the 19th century a very famous landscape architect, Frederick Law Olmsted, worked across the whole United States, and created a whole series of great urban parks. These parks are used by everyone, people love them, and have loved them since the day they were built until now. That is the kind of project which is really important to design: projects that meet present needs of the whole local population and that can be passed down to the future. We should also design for the people who will come after we are gone.

WU: Nowadays there are many designers in the field of landscape architecture who develop projects for "green infrastructure"—greening is a fashionable word—such as designing in a city a river valley which imitates a river flowing through flatlands in order to achieve the same sustainability that would apply in this natural environment. Yet your designs do not imitate natural forms, they introduce large figurative forms, the greatly magnified outlines of some living species, so can you explain how they can nevertheless be self sustainable. Could you for instance turn to Petaluma, your latest project for a water treatment plant which just opened last year, and has received much attention?

JOHANSON: Well, river flow is not a technical process, water treatment is. There are natural processes that shape rivers through time in any environment; this is not the case for water treatment. Yet cities need to clean the water used by its inhabitants and industries, and its storm water, and this water treatment has to achieve a form of its own. At Petaluma, the design follows the image of a little salt-marsh harvest mouse, an endangered species that lives on the site. That is the third image that was proposed for that design, because during the study process some circumstances led to changes in the "treatment train", the technical and natural steps followed in processing the sewage and sullied waters. To make it sustainable, first we wanted to achieve as low energy use as possible. So the water mostly flows by gravity throughout the plant. There is only one instance when we are pumping water; otherwise it is always flowing downhill. Second we were concerned about the conservation of building materials. There could have been a lot more hardware, a lot more technical equipments out there to process the sewage. Third, it is built to stay close to the earth. I tend to prefer things which are really simple, really basic, and I designed earthen ponds. They are really simple to build, inexpensive and really easy to repair. But there is more to self sustainability of the whole environment of such a project, since that area of California has experienced some big earthquakes, and a dam failure can wreak havoc downstream. Earthen ponds are good in earthquakes, and these ponds are designed so that the worst thing that can happen is that if there is an earthquake, the water-already clean enough for agricultural use or release in nature—would go back into the river a few hundred meters away, just flow down and go down to the river.

So, basically it is the weaving together of the technical process of the treatment train, the environmental and landscape design which insure the self sustainability of the project, not the form into which they are cast.

WU: You have explained the technical and environmental dimension of the project; can you say what the art dimension is?

JOHANSON: Art is not an added frill; it is integral to the design. Let me give just one example. There is one part of the project devoted to treating storm water, entering the site from the local highway and a neighboring business center. This is contaminated water containing oil pollutants and chemical fertilizer, which should not be allowed to run directly to the Petaluma River. I have designed a separate filtering process, driving water through a series of small planted ponds, in the overall shape of a morning glory. It forms a compartment of different water plants in several colors forming with a pool in the middle the figure of a morning glory flower. Now the aesthetic part of course is the color and the figure; the functional part is the choice of filtering plants each selected to treat a specific pollutant present in this water. So, if you look at the image of this flower from the top of the berm of the neighboring pond, you get on the same footprint the art figure, the colors, the plants and the technical process.

WU: How would you expect visitors to appreciate the project in Petaluma? Should they gain an understanding of the design of the water treatment train? Or turn their attention to the protection of the Salt Marsh Harvest Mouse? Or do you expect something else?

JOHANSON: I think different people look at things in different ways. So if you interested in wildlife, you probably come there and see the wildlife; if you are a hiker and you are interested in parks and trails, you might jump on a bicycle or juggle along the trail; if you are a botanist, there are wonderful plants out there that people appreciate. I am not really sure how people first look at it. It can be quite unexpected.

Not all the art is in place yet; we still have to add colored sand in the finishing ponds. After this is done, you will get the color reflecting through the water. This is very important because it will focus attention on the flow of water, and the ever changing appearance of water. It is also a deeply visual phenomenon, and you might expect this to be a necessary trigger for an aesthetic appreciation of otherwise rather technical water treatment ponds. Yet, upon my last visit, I could see that there were several painters who had planted their easels and engaged in plein-air painting on the bank of one of them. Claude Monet had planted a nymphea in his garden to enable him to paint water lilies in a Japanese setting; some inhabitants of Petaluma have seen the islands in these ponds as places that stimulate their own sense of nature. These islands are part of the engineering since they are directing the flow of water from one finishing pond to the next. So they are part of the technical infrastructure, but they are also designed to attract wildlife, to provide nesting areas for birds. And, I just think that at the moral level, it is really important, every time that we build something, to remember that we encroach upon land for wildlife, and we have a moral obligation to make up for this inconvenience to wild life. That is one of the reasons behind my selection of plants. I choose them to create new habitats and nesting sites for all sorts of birds and animals. These animals are not captive; they come because they are attracted to the vegetation, to the substrate, to the presence of other animal life upon which they prey. They may also provide entertainment to people who show their child an otter, or a fox for them to learn about wildlife beyond what they see in their yard. The more reflexive will see beyond this level. They will see that when designing for wildlife this project did not start with what you see, but down to the level of the microbes, the much tinier ecosystems of microbial plants and animals, which are actually responsible for a great deal of the water quality and treatment. They will reach through their personal approach a new level of ethical engagement.

WU: You see landscape appreciation as a process that develops over time: Landscape architecture should give pleasure, then invite aesthetic appreciation of wildlife and then stimulate ethical engagement in environmental issues. Does that imply a new design practice as well?

JOHANSON: I think landscape architects have a huge contribution to make, because they are so involved in constructing the urban environment. Let me stress the role of time. Once a certain type of landscape starts being built, we train our eyes to like it. One of the conflicts in the United States between natural landscape and manicured landscape, the kind all landscape architects used to do around housing or public buildings, is that people wanted to see a lawn; they wanted to see cut grass, that would eventually call for the use of pesticides, which would flow into the water with all ensuing problems. When confronted with a natural landscape they felt it was just a field, it looked like weeds it did not look like somebody was taking care of the landscape. However this may be changing as the years go by. So, landscape architects are responsible for training our vision in the sustainability direction and to do that they need to think about planting not so much in visual terms, just to please the eyes, but in terms of the contemporary world, where the population is expanding and all the resources are being stretched. I think that they should engage in the development of multi-functional, multi-purpose, public infrastructures which provide a new and important support to the development of wildlife at the same time that they efficiently provide a self sustainable service.

This implies engaging in cooperation with large numbers of people in any design project. And even though I know it is difficult, and even though it takes much time, and even though nobody gets one hundred-per cent of what they want, I think that working together achieves a better project than you would by working individually.

I can only speak for myself. I like to do new functional projects. Each project becomes a kind of model for a certain type of situation, just like the reclamation project I am doing for a really toxic coal mine in Pennsylvania now. This is a place where you cannot grow anything, where the water just disappears below the surface of the earth. I want to bring that land back, make it productive and beautiful and useful. I want people to be able to go there, and to see the water and to swim in the river again, which they cannot do right now. Those are the kinds of projects that I want to do, even though I do not do many of them. Each time I select a project, I try to address some problems that are really significant. That is what I shall keep doing, as long as I can. (Translated by Yunsheng LI, Proofread by Xin WU)

埃里克·董特
——当代美学与欧洲造园传统

埃里克·董特是一位欧洲设计师,工作室在布鲁塞尔。由平面设计师转为景观设计师,董特的作品将抽象的造型与本土生物群落和欧洲花园设计传统巧妙结合,在业界广受赞誉。值得称道的是,这位景观设计师所创作的作品中包括了众多的花园,而作为一名现代艺术家,他还参与了一些历史性花园的复原。他用艺术的手法对景观进行创造性的设计,并融合欧洲传统和当代美学,在历史与现代、本土性与国际性之间成功地搭设起了沟通的桥梁。对他而言景观设计应该,也可以包容古典传统和新型美学,因为两者都基于对文化的理解,是对历史的重新解读和对人类未涉足的各种可能性的探索。董特的兴趣十分广泛,从私人花园到历史景观、公共环境、园艺、平面艺术、家具设计和视频装置等。为人们的"未来之梦"(a dream of the future)而设计,他的作品回应了各类业主的需求:从商业大亨到艺术品收藏家,再到雅皮士;从怀有强烈历史感的慈善家,到都市时尚人士,再到弱势移民群体。董特大部分时间都在比利时、荷兰和法国工作。过去10年间,他受邀参与了一系列国际项目的设计工作。他也经常应邀到设计院校演讲,并是英、法、荷三语版的《董特:花园和秘密景观》(DHONT: Gardens, Hidden Landscapes)(Ludion, 2001)一书中的话题人物。

我将董特的努力定位于欧洲独有的文化和艺术背景之下。景园美学是18世纪欧洲众多美学争论的中心。到了19世纪,由于折中主义和仿古风格的兴起,艺术家不再热衷于创造新的花园形式,景园美学才逐渐淡出争议的焦点。20 世纪现代艺术蓬勃发展,出现了印象派、野兽派和立体派等运动,而现代造园艺术却未能在欧洲有实质性的发展,这种停滞状态一直持续到第二次世界大战。因此,到了 20 世纪中叶,欧洲的景观师及其客户已没有任何美学传统可以参考借鉴。直到战后,法国的贝尔纳·拉素斯 (Bernard Lassus)、苏格兰的伊恩·汉密尔顿·芬莱 (Ian Hamilton Finlay) 等一批有远见的艺术家们才开始重建艺术与花园之间的联系。20 世纪 60 年代兴盛的"情境主义国际"(Situationist International) 引发了大规模学术争鸣,并在荷兰、斯堪的那维亚半岛和德国掀起了绿色运动,埃里克·董特从中获得了大量创作灵感,特别是"情境主义国际"的追随者、荷兰艺术家路易·纪尧姆·勒华 (Louis Guillaume Le Roy) 的实验艺术,让董特获益匪浅。勒华因提倡新型的生活态度,反对无限制消费及对自然的破坏,在国际上声名斐然。他倡导自发的自然发展和人类与自然的日常接触(机械和化学干预的活动除外)二者之间的新型平衡关系,并呼吁将自然界的自我组织能力看作是人类自身美学享受的源泉。勒华的观点显然与同时期的美国人伊恩·麦克哈格 (Ian McHarg) 大相径庭。董特发展并改变了勒华的美学观点,呼吁人们关注人类艺术干预和自然界自我组织之间的平衡。这一目标更为复杂,需要艺术创作与自然环境和谐共处。现代主义将二分论的美学引入艺术,在现代性和历史性上没有丝毫妥协的余地。尽管如此,景观设计似乎仍处于"白板论"(Tabula Rasa) 的魔咒之下,强求在没有传统束缚的净土之上进行现代主义设计。而埃里克·董特的作品则证明,当代景观设计完全可以摆脱这种困境,重建自我。他的作品在达到这一目标的同时,也推动了园林艺术在西方的新生。

吴欣：几年前我去了趟比利时，见过你设计的作品，印象特别深刻。今天我重点想了解一下在这些作品背后，你的思维过程与方式，特别是你的设计理念。你成功地把当代设计引入了历史性的景园之中，我们就从这一点开始吧，因为这在中国是个很重要的问题。中国现在所面临的历史性地标和现代化发展之间的冲突和欧洲的情况是很相似的。

埃里克·董特（以下简称董特）：我在学生时代就对当代艺术、历史以及文化在历史过程中的转变十分感兴趣。当前，从事历史园林修复的景观设计师都将太多的精力关注在历史遗产的修复上。他们想极力重现历史性花园当年鼎盛时期的外貌，哪怕大多数时候并不是很成功。然而，需要一提的是，就算这些园林在法国、英国、比利时、荷兰等国家被奉为国家遗产的一部分，我们现在的生活和思维方式仍跟400年前的17世纪有很大的不同。因此，过去人们的生活方式和诠释环境的方式与现今已大不相同，中间隔着不可逾越的鸿沟。过去就像是另一个国家，说着不同的语言，今天的人需要翻译才能够对其理解。对于园林和景观来说也是如此。

然而，对于景观设计来说，存在于自然世界以及人类的尺度和比例之中的逻辑是不会变的。就是说，只要能在所创建的环境里维持一种平衡，我们就可以在现代的环境中建造历史性花园。尺寸、材料和比例都必须经过考虑。我们还可以选用某一特定历史时期传统的植被材料来重建花园，哪怕设计是截然不同的。每一处历史园林都有其历史价值的积淀，每一处都是独一无二的。没有放之四海而皆准的万灵方案，而长期以来形成的价值观也是不能改变的。设计师可以重新启用这些价值标准，并对现代艺术做出贡献。当今的景观设计师在创作时希望对现代艺术有所贡献，但经常会急功近利，太急切地想展示出设计师的自我。我觉得我们应该更谦虚一些，应更加尊重周围的环境。只有当我们同时在微观世界和宏观世界里意识到自然的存在，我们才能成功修复历史园林，并将其价值融入现代文化当中。作为景观设计师，对我来说，这是我们这个时代最基础的两种价值观。

花园得以修复之后，还必须注意维护和管理方面的细节，这一点也很重要。我们可以有效维护园林，使它们的样子与修建之初保持不变，但如果种植了树木，却又要花费大量精力去阻止其生长，就显得有些不切实际了。在我看来，协调好与自然的关系，建立起良好的平衡才更为重要。比如，我们可以只对项目中的一半进行设计，而把另一半则留给自然去做功。通过这种简单的方法，你便可以在人类、植物与野生生物之间达到良好的平衡。越是保留其原始的荒野状态，越是有更多的鸟类飞来驻足或筑巢。作为设计师，我对自己的身份一直都很清醒。种树这种行为改变的景观将会与许许多多方面都相关联，我们必须对其未来和生长状况负责任。

吴欣：这让我联想到加斯比克（Gaasbeek）的一个项目。幸运的是，你的现代设计得到了法兰德斯遗产学院（Flemish Heritage Institute）遗产研究员赫曼·范·登·博斯（Herman van den Bossch）的支持。但大多数情况下，许多人都认为历史和现代难以融合在一起。你觉得这里面会涉及美学的问题吗？

董特：的确，这确实是个问题。美学价值观在不断改变，它反映的是艺术的创新。我这里讨论的是设计，也就是美学价值观。包括很多设计师在内的许多人都相信，创新就意味着反传统。而我却认为，在借鉴环保与生态运动中倡导美学价值的同时，保持传统的花园设计也是可能的。通过这样的方式，我们不但可以借鉴园林传统，同时还能让花园与自然达到平衡。可以在现代设计中融入传统之美，比如十六七世纪的园林会有金字塔形、球形、动物以及方盒形状的灌木，以创造出古典的修剪效果。用现代的方式也可以达到这种灌木修剪效果①，但是

---

① "灌木修剪法"（topiary）是个现代词汇，来源于古老的拉丁单词"topiarius"，是"园丁"的意思。在罗马帝国初期，正好是公元纪年开始之前，随着花园在罗马成为一种时尚，这个单词开始出现在拉丁词汇里。它来源于希腊单词"topia"（原意是"场所"），到了罗马，它被借指壁画中组成景观的各个不同的场所。因此罗马的园丁被比作是画家，虽然画家是用颜料绘制出真实场所的错觉。然而，在现代英语中，"topiary"却用来描述园丁创造几何形状或形象图案（如动物和人物的简单轮廓，甚至是军队或是围猎野生动物的一群猎人）的方法。16世纪时，灌木修剪法在北欧十分盛行。但是到了巴洛克时代，上流社会对这种做法十分不屑。直到19、20世纪折中主义兴起之时，人们才逐渐对古老的灌木修剪法重新产生兴趣。最壮观的灌木修剪作品出现在19世纪末、20世纪初的英国，尤其是帕克伍德庄园（Packwood House）

要用更为抽象的形状,因为相比形象图案来说,我们喜欢抽象图形更多一些。即是说,我们可以用一种很不同的设计来诠释传统。我们还可以关注诸如光线在植物上投射的影像之类的事情。不过,在现代景观设计中,让这些历史园林的所有者理解到保持这些价值标准的可能性也很重要。并非总要创造出新的事物;可以先保持已有的,然后再加入现代的元素——这或许就足够了。

吴欣:你提到了"转译"。你能不能再进一步解释一下,在具体的项目实践中,比如在加斯比克的项目,你是如何转译的?都转译了些什么?

董特:那座城市 1602 年就有了,而花园是现在建的。这座历史古镇里有一座文艺复兴风格的园子。这就意味着花园里有修剪齐整的绿篱、遍处的花草,以及矩形的花床。迷宫是文艺复兴时代很重要的特点,它让园子更有趣味,这类园子叫做"闲趣园"(Plaisaunce)。因此我们提议用迷宫这种元素将文艺复兴风格带到园子里来。这种转译必须符合历史,但设计必须是现代的,这样一来,在迷宫里散步才能符合现代的思维方式。因此,迷宫不能按照传统的方式设计。我们设计出一个抽象的迷宫,有些设计细节,比如玫瑰花会一路引导着游客,直至在自然里完全放松。我们修建的花园里就有这些抽象的概念。花园是 1992 年建的,在 20 年后的今天来看,它仍然维护得很好,而且还在与时俱进。园内的植物和树木生长状态都很好,将来会长得很好,保证这一点很重要。景观设计师必须想象得出花园未来会变成什么样子,当然不可能准确地预知以后会发生什么事情,所以 20 年后再回到这个地方会是种很愉悦的体验,因为你依旧能感受到当初设计的力量,以及自然的美妙和花园的成长。设计的时候必须很谨慎,因为如果使用太多当代的材料,或是太多会很快变质的材料,那整个项目在未来

里的"登山宝训"(the Sermon on the Mount),用修剪过的欧洲紫杉(*Taxus baccata*)抽象而壮观地刻画出众教徒听取基督布道(公元 30 年)的场景。从下文的访谈中我们可以了解到,埃里克·董特这位当代罕见的园林艺术家正是利用欧洲紫杉良好的塑性,努力重现修剪艺术的辉煌,而使用的植物正是帕克伍德庄园里种植的欧洲紫杉。

就有可能被毁掉。这也是为什么我喜欢用天然的石材而不爱用水泥的原因,因为石材的老化现象不会太严重。当然,出于结构上的考虑,也应该去利用一些现代的材料。

吴欣:谈到迷宫设计,你曾经这么说过:"迷宫是什么?它不是指引着你、让你面对自己的趣味之旅吗?"这是否就是你刚才所说到的园林的抽象概念呢?

董特:没错。设计花园时,我们一直在做同样的努力,努力不让建筑脱离自然而存在。居住者对于房屋有一种深厚的心理依赖,而我们的目标是让他们远离房屋,亲近自然。花园能引起人们的好奇心,让人们注意到自然,从而发现周围的环境,并爱上自然环境。正是花园让我们注意到自然的事物,让我们的心境平静下来。这一点在现代越来越重要了。最近我们在荷兰设计了一座秘密花园,人们可以在园子里放松心情、进入梦乡、观赏身边的植物,之后慢慢地注意到其中的美感与乐趣。

吴欣:跟你交谈总是心情很愉快。提到花园时,你会用"爱、尊重、快乐、入梦"等词汇,这些概念帮助我们了解到,花园是有多层情感空间的地方,而不仅仅是建筑空间。

董特:建筑的本质在于其内容与功能。建筑师必须要创造结构,并在某一地方搭建起这种结构。而我觉得景观设计师关注的都是其他方面的事情。景观设计的根本层面并不是为了解决某种迫切需求——他们往往在第二层面上才考虑这些事情——景观设计师更关注外观、气息和体会。事实上,这里的第二层面更需要我们关注周围的环境。景观设计所处理的是一种梦境般的自然。我们是那些制订方案、提出梦想的人,当然,我们也种植物,这些植物也会生长,而梦想终有一天也会实现。许多人都想要个花园,但他们的耐心往往不够,因此设计稿上再美丽的园子,其现场也往往很糟糕,因为他们忽视了时间的演化。就是说,我们应该更加重视时间对于美景的影响。设计一开始往往是一种绘制梦想的方式,或者是生活的方式、观察的方式。但客户看待设计稿的眼光与设计师不同。我们会邀请他们用不同的视角去观察，有时候在局限的空间里，有时候在开放的空间里……当然这取决于地理位置。

吴欣：把设计花园比作是实现梦想，这个比喻很美妙！这样一来，随着时间的推移，设计师会更加注重花园的发展。种树的时候，我们会考虑这些树以后将怎样生长，而环境对于树木的生长又会有什么样的反应。景观设计师的责任也因此变大了，难道不是么？

董特：当然。有必要好好观察古老的花园，因为在那里我们可以发现长时间以来各种事物的变化。在历史性花园里，我们能够看到过去的人是如何使自己的行为迎合周围环境的。我们确确实实是在向先辈学习。而有了这些知识，我们可以进一步在现代社会里进行诠释，也清楚地了解到，我们现在的生活方式与过去是不同的。但不管怎么不同，人类对美的欣赏是相同的，因此人们始终在追求美给我们的生活所带来的快乐。

吴欣：当你审视一个具有历史意义的地方时，你不但关注其文化遗产，还会注意到那里的自然遗产。

董特：很对。我得强调一下文化和自然遗产之间的关系。通常，我们看园林时关注的是主轴线①。所以当我们在园林里观赏的时候，我们会朝着主方向看过去，而在这个方向我们也会感受到一种力量，一种建筑的力量，一种展示的力量。不过，对我来说，最有意思的是垂直的那条轴线，因为那是花园浪漫之所在，是看待这个地方的另一种视角。路易十四曾写过关于解释如何介绍凡尔赛宫的花

---

① 从意大利文艺复兴时代开始，欧洲的闲趣花园都尽可能根据其所在位置进行布局，并围绕垂直于主建筑物、城堡或宫殿正面的轴线以对称的方式展开。从沃子爵城堡（Vaux le Vicomte）和凡尔赛宫（Versailles）开始，人们越来越关注这条轴线，欧洲园林也都遵照勒·诺特尔（Andné Le Notre）所创造的方式进行设计，并使这条轴线趋于无限远。埃里克·董特这里所说的正是这条轴线。

园[①]的一些文字。他写的是如何观赏园林、喷泉和宫殿,以获得最大的享受。用这种方式观赏,我们能不断发现不同地方的景致和乐趣。这些介绍是为了让访客看到宫殿的每一处地方,因为每个地方都已证明了其自身的逻辑,而不需要那些额外的解释。这正是景观设计和园林所展示出来的美,它遵循的是自然的逻辑。以自然的逻辑进行设计的话,其作品能达到更好的效果,也会更美。

吴欣:我记得你曾跟我的学生介绍过如何观赏伯马佐花园(Bomarzo)[②],那很有启发性。现在我们来谈谈你的设计方法。首先,你是如何从平面设计转向景观设计的?

董特:那纯粹是机缘巧合。我的一个朋友有个苗圃,种着果树、观赏植物和花卉。有一次见面时,他注意到我对花园和景观设计很感兴趣,因此向我介绍了许多相关的内容。平面学校更像是个印刷学校,讲授书籍和文字设计的知识,帮助我从不同的角度看待设计,我也更关注比例的问题。植物的世界是能够作研究并从中学习的地方,至今我仍然每天都在学习,因为所处的每一个环境都是不同的。

吴欣:很多人对你的园林创作草图十分着迷。你能谈一谈这些草图在你的

[①] 现存路易十四亲手写的几页文字阐述了应该如何向国外使者介绍凡尔赛宫的花园,历史学家已对此进行研究并予以公开出版。参见西蒙尼·胡格(Simone Hoog)所著《路易十四:如何展示凡尔赛宫的花园》(*Louis XIV: A Way to Show the Gardens of Versailles*)(巴黎,1982),还有最近托马斯·F·海汀(Thomas F. Hedin)和罗伯特·W·伯格(Robert W. Berger)的合著《路易十四指引下凡尔赛宫花园的外交之旅》(*Diplomatic Tours in the Gardens of Versailles under Louis XIV*)(费城,2008 年)。

[②] 伯马佐花园(The Park of Bomarzo)于 1552 年由维奇诺·奥尔西尼(Vicino Orsini)建造,而奥尔西尼又因其妻子与法尔尼斯(Farnese)家族有姻亲关系。据说 16 世纪著名的意大利考古学家皮诺·利戈里奥(Pirro Ligorio)是这座园子的设计者,园内的雕塑精美绝伦。伯马佐花园与意大利文艺复兴时期的其他园林有很大的不同,并在 19 世纪逐渐为世人所遗忘。直到 20 世纪前几十年,超现实主义诗人和艺术家才让它重新名扬四海,从此声名远播,并引发了众多争议,也激发了许多艺术家的创作灵感。

设计工作中所发挥的作用吗?

董特:画草图是思考的一种方式,是探索空间如何定义、行动,如何定位到方案中的方式,是在空间里漫步的一种想象方式,也是为了解决问题的一种想象方式。没有任何一位建筑师或景观设计师可以很轻易地实现某些要求。我觉得在景观设计中,在详细研究各种物质局限之后,通过画草图的方式能更加自由地想象,虽然脑海里仍然想着各种各样的局限,但画草图时可以暂时不予考虑。画草图能够把自己从技术和功能的需求中解放出来,进而发现园林中其他不那么物质的方面。之后,草图所带来的可能性会接受既有局限的考验,而整个项目也会因此找出解决方案。

第二步是寻找草图灵感和现实局限之间的平衡。草图被我用作远离日常实践中的思维习惯,并在大脑里不考虑任何项目的情况下进行自我训练。举个例子,乘飞机的时候就可以把大脑里的想法画出来。这不是无聊的涂鸦,这些草图对我来说很重要,因为草图里的想法并没有关联到任何项目,而只是在训练我自由思考植物与空间的能力。这对我来说尤其重要,这是需要继续加强的一种横向思维方式①。从实践的角度说,设计师的积极活跃的生活应包括自由思维和考虑现实局限与问题两部分。为了使自己的作品更加优秀,设计师应该进行这种自我培训,而画草图则是实现内在升华、增加内在积淀修养的途径。如此一来,设计过程中如果遇到了什么难题,就可以根据自我训练获得的内在积淀,提出创新的设计方案。

吴欣:你这种观点一定能让现今依赖计算机的年轻设计师深受启发。你的许多设计草图让我联想到抽象绘画。从抽象绘图转化为景观设计,这是不是你在

---

① "横向思维" (lateral thinking) 是爱德华·德·波诺 (Edward de Bono) 在其1967年出版的第一本解决问题方面的书——《新想法:运用横向思维》(*New Think: The Use of Lateral Thinking*) 中创造出的词组。此后出现过大量深入浅出的书,全都在探讨如何通过间接的方法锻炼创造性思维的能力。归纳在创造性思维过程中发挥的作用这一严肃课题通常只是学者们讨论的话题,但这些书的作者为此类学术讨论提供了一个通俗的版本。在这里,埃里克·董特并不是简单地重复书中有关创造性思维的方法,而是在表达他自己的方法。

画草图的过程中所追求的发展风格呢？

董特：我觉得这更像是把景观设计当成一种艺术形式。每个人都是居住环境中的小孩，看着新奇的当代建筑或绘画的时候，我们对熟悉的生活环境会有全新的思考方式。通过解读别人的艺术作品，用以解决当今世界中的难题，这种转换过程很有意思。但这还不是艺术，而是一种观察方式。艺术家看重的是艺术品的创作，而对于景观设计师来说，重要的是看懂艺术品，因为这让你能够看懂艺术家在现实世界和美学面前是如何取舍的，也使你能够自由思考，远离职业思维。还能通过这种方式，激发创作灵感。不过，我们也不要忘记，每一种艺术都有其自身的媒介。景园艺术和景观设计行业所关注的是自然里的空间，从古到今都是如此。

吴欣：能不能说你的景观设计就是关于自然空间的一种抽象艺术？

董特：可以，我觉得这么说很准确。这也是我追求的目标。

吴欣：做园林规划时，各不同部位之间的张力在其图形构成中十分重要。你是如何将这种张力转化为建筑空间的？

董特：我们可以在确定植物在花园所种植的空间位置时转化这种张力，直至达到一种平衡。如果设计的元素太多，或是把太多内容塞进某一个空间里，这种张力将会分崩离析。而在适当的时候停下来是很重要的。我们把能量带入一个空间，也要知道什么时候应该停下来，因为太多内容反而不能达到理想的效果。这一切都是平衡的问题。这解释起来或许要花费很多笔墨，但这其实就是某一个特定的"点"：可能是与空间相关的树木的年龄，或者是文化遗产，或者是环境的作用……比如，对山谷里的花园来说，山谷景观就显得十分重要，因为这将影响到花园访客的感受。张力也会随着季节和天气的变化而改变，因为张力跟光线、风、雨、花草和鸟类的存在都有关系。这些都需要我们通过观察和实践不断进行学习。

吴欣：那雕塑形式呢？它在你的设计中处于什么样的位置？

董特：你可以雕刻出花园的空间，就像野口勇（Isamu Noguchi）在"加州剧本"（California Scenario）公园里所做的那样[①]。任何一处景观中都有开放空间、阴影遮挡和一定的容纳体积，但另一方面，它也是一处古迹。我们还可以在某个地方整合新的雕塑景观，就像罗伯特·史密森（Robert Smithson）在"螺旋防波堤"（Spiral Jetty）[②]所做的那样。它改变了我们对景观的看法。20世纪六七十年代大地艺术家创造了大量类似的作品。那已经是四五十年前的事情了，而今时与往日不同。他们的创作在当时很重要，而现在我们回过头去看，再把大地艺术和西方更伟大的景观传统联系起来，会觉得很有启发意义。

吴欣：在花园里你是如何直接处理雕塑的？

董特：最近我刚在美国做完一个项目，主题是客户的当代雕塑收藏。这就出现了一个难题，因为我们不希望各个不同的雕塑间相互干扰。这要求我们对景观进行改造，把我们不想看到的隐藏起来，从而创造出另一种视觉环境。修建这座雕塑公园时，我把园子划分成不同的区域，以使各个雕塑都能保持自己的视觉完整性。通过这种方式，我们对景观进行造型，激发出灵感，也会产生一种亲密感，或是在周围环境中营造出不同的气氛。通过这些不同的手段，景观和雕塑可以面

[①] "加州剧本"（California Scenario）是一座公共花园，1979年由开发商、慈善家亨利·T·赛格斯特姆（Henry T. Segerstrom）在加州科斯塔梅萨市（Costa Mesa）修建而成。赛格斯特姆委托野口先生设计这座公共花园，本意要通过设计一座公共绿地来提升建在其间的两座写字楼，而这片土地原本是种植利马豆的田野。但野口并没有按照要求进行设计，而是创造了一处简单的岩石广场，仅有零星绿地点缀其间。野口博物馆（Noguchi Museum）这样写道，创作者最早对这个项目的构思是"加州的抽象隐喻，包括锯齿状山脉、沙漠和树丛。除了种植红杉、仙人掌等其他当地植物外，公园还包括了创作者设计出的一系列个性元素，以强调加州突出的特点"。引文摘自2010年野口博物馆在纽约市举办"加州剧本"专题画展时发布的公告。

[②] "螺旋防波堤"（Spiral Jetty）是罗伯特·史密森最著名的作品。这项土木结构修建于犹他州大盐湖（Great Salt Lake）荒凉的岸上。它不同于在一般都市博物馆中展出的作品，其证明了在人迹罕至的地区，利用填埋垃圾这种不那么光彩的材料进行艺术创作的可能性。不管从哪个角度看，它都是纽约艺术世界的复兴宣言。

对面进行对话。如果空间不够大，可以利用绿篱和修剪过的树木来达到类似的效果，比如可以剪出岩石的形状。我通常用欧洲紫杉，因为它既有树的自然体态，又有人工修剪出的形状特点。

吴欣：我想请你谈谈如何从自己的想法中有所收获。现今的许多设计师都不会、甚至不愿意回过头再去看看已经完成的项目。你会重访旧项目，你的团队也负责这些项目的维护工作。你甚至还创造出一种叫作"三种变体"（Three Variations）的工作模式，以不同的媒介探索同一个想法。你能给我们讲讲这种做法吗？

董特：回顾自己的设计其实是回过头去看自我的发展历程。重访旧项目时，可以看到哪些方面做好了，哪些地方犯了错，因此从自己的经验中学到新的东西。我们对某个花园的想法或方案只是针对某一处具体的项目设计，不能复制到其他的项目中去。但在制造出大量的变体之后，我们便会开始自我复制，这时候也就没有什么创造性可言了。这就是"三种变体"概念的来源。

对于设计师来说，重新思考自己的作品，从自身经历中学习，并从一些绝妙的想法中衍生推出几种变体设计，是很棒的做法。但最重要的是要限制对自己作品的引用，并要与时俱进。如果能在制造变体的同时不断进步，就能培养自己创造新方案、解决不同设计难题的能力。

吴欣：你的意思是不是说这是发展个人设计风格的一种方法，就像你画草图做法一样？

董特：我并没有在追求发展出一种风格，而是想形成一种我个人的"签名"特征，也就是个人创造设计方案、做研究和分析问题的方式。这就是我所说的特征。风格说的是另一回事。风格和时尚相关，是一种强化效果，会限制创造的潜能。与之相反的是，设计师应该保持自由，不断鞭策自己，开拓自己未曾涉足的领域。

吴欣：这么说，作为设计师而言，自由的创造性思维对你来说相当重要。

董特：非常对。尽管在实际项目中总会有实际的问题和局限，从而带来各种各样的压力，但这也是好事。

吴欣：你和不同的艺术家们合作时运用的媒介也千差万别，比如有的用现代音乐，有的用舞蹈。你也还曾经设计过家具。

董特：我对舞蹈这种元素进行了琢磨思索。花园之美会让人忍不住跳起舞来，这是一种不由自主的行为。而家具其实是设计的一部分：如何才能让你设计的椅子用上10年也不坏、柜子就算在室外放上30年也还能用？我对比利时伟大的传统手工艺进行了研究。你观察音乐、家具、设计和建筑这些不同的内容时，会发现所有元素都会互相呼应。这是发现和探索的一种方式。

吴欣：景观设计师提到研究的时候，通常会从生态学家、植物学家等科学家那里寻找某些有用的研究信息。这跟你的研究大不相同，你是为了提高自己的设计能力而亲自做的研究。当前你主要的研究课题有哪些呢？

董特：目前我正在研究花园形式的传统，特别是传统的树木修剪法。树木可能修剪成哪些形状呢？可以是立方体、棱柱体、金字塔形，或是任何几何形状，同时也可以是某种生物的或是塑料的形状。我在几个不同的项目中进行探索，我得说，有时候会需要花费工夫去琢磨，但这种探索是相当有意思的。我也在研究人们散步时走几步之后会向左看、走几步之后会向右看这些事。同时也在研究做花园设计时，哪些事情应该怎么做、哪些事情不应该做。通过设计为场地赋予内涵，这一点很重要。比如说给周围的空间添上最流行的色彩，使视线聚焦、强调游客观赏植物和自然时的视觉体验等。但被赋予的这些内容可能会产生不良的效果，甚至会使人为干预和自然的自由发挥之间失去平衡。保持这种微妙的平衡也是景观设计师的工作职责。

吴欣：你提到的景观设计中该做和不该做的两难，这让我联想起你对荷兰艺术家路易·纪尧姆·勒华的作品的赞美，特别是他在荷兰小镇米尔丹(Ｍｉｌｄａｍ)

附近创作的"生态教堂"(Eco-Cathedral)①。

董特：勒华这位艺术家十分关注战后的生态运动。然而，至少可以这样讲，生态运动对于美学的立场并不是很明确。它虽然能够解释如何让景观中的某些植被生存并维持下去，但在创造性设计方面却十分薄弱。勒华写过一本书，叫《关掉自然，迎接自然》(Switching off Nature, Switching in Nature)，给我留下极深的印象。我第一次听说通过废弃物利用进行景观建造，利用拆迁工地的砖石修建新型花园，而不是将废弃物扔进垃圾堆，制造环境污染。生态教堂是他的杰出作品，通过回收再利用废弃物，堆砌出新的纪念建筑。更重要的是，他解

---

① 参见斯帖·布科马（Esther Boukema）与吉多·范·奥佛比克（Guido Van Overbeek）合编的《自然与文化的融合——路易·纪尧姆·勒华的生态教堂》(Nature Culture Fusion — Louis G. Le Roy's Eco-cathedral) 一书中路易·纪尧姆·勒华的引文。摄影：菲利普·维莱斯·麦克英泰尔。附加文字：皮特·沃拉德、哈根·罗森海因里希、文森特·范·罗赛姆。该书已由荷兰建筑学院以英、荷双语出版，在专门销售阿姆斯特丹艺术家著作的Boekie Woekie书店有售。网址：www.boekiewoekie.com。

路易·纪尧姆·勒华是法国胡格诺（Hughenot）家族在荷兰的后裔，出生于1924年，目前仍然居住在荷兰希仁温市（Heerenween）橘木地区（Orangewood）。20世纪六七十年代，他的早期作品批评人们不够关注自然及其自我组织的能力，因而在德国和荷兰产生了很大的反响。在参与了肯尼迪拉恩市（Kennedylaan）双车道公路中间1km长、15m宽的分隔绿地改造项目之后，他在德国接到了许多项目，但都以失败告终，因为这些项目都注重人与自然的互动，这让"资深生态学家"很不高兴，而他所坚持的"自然与人类自由活动"的理念也引起了当地行政部门的不满，因为政府官员认为他们有义务在自己权力范围内对一切事物施加法规与限制。勒华的思想深受亨利·博格森（Henri Bergson）和盖·德博尔（Guy Debord）这两位法国哲学家的启发，同时也在艺术家康斯坦特·纽文华（Constant Nieuwenhuys）及其1957年创立的"情境主义国际"（Situationist International）身上获得了创作灵感。路易·纪尧姆·勒华的一个孙子最近在他的网站上写道："我爷爷是位艺术家、哲学家、作家、画家和摄影家。"这话说得很对。而事实上，诺贝尔生物化学奖得主伊利亚·普里果金（Ilya Prigogine）关于有机自然生产顺序方面的写作也对路易·纪尧姆·勒华思维的发展产生了十分重要的影响。大约30年前，他在米尔丹（Mildam）购买了一块田地，并在上面开始进行他的个人项目——生态教堂（Eco-Cathedral）。所有建筑材料（岩石和石块）从希仁温市运往他购买的田地上，用于建造一座"教堂"。简而言之，他的哲学是：我们应该让这些废弃的碎石堆与自然进行接触，用双手搭建出能够表达我们自己的生活方式的结构，然后让自然来组织这个结构周围的生命。自然将会证明其自身的组织能力，并且展现超出人类规划范围的结构。发现这些结构的同时，也将会发现这种景观之美。他在这片田地上只身修建了各种雕塑、小径和台阶等构件。此后自然接过手，植物不断生长并逐渐覆盖这些结构。他立誓说，人与自然的互动必须在这项工程中持续进行到公元3000年。

释了哪些事情该怎么做，哪些不应该做，以及如何平衡"有为"和"无为"之间的关系。一方面，他利用废弃物搭建起的辅助景观构筑物允许附近居民作日常取道通行之用，并为过往路人提供简易的便利设施；另一方面，他任由自然在这些辅助构筑物里施展才华，自由生长。他在人为活动和尊重自然之间创造出了平衡。最重要的是，这是一种互动的关系，在人与自然自由的互动过程中激发景观的美学鉴赏体验。在我看来，这种方法既尊重了自然，又不会削弱设计的分量。

吴欣：我注意到你也在几个设计项目中利用了废弃物，甚至在为一对新潮的年轻夫妇设计的项目中，你也成功说服他们，接受这种极富挑战性的新型美学。

董特：大体上讲，我的客户都很有教养。但为了说服别人，你得有想法才行。设计师必须能够传达自己的想法。如果无法传达，这些想法终究只能是文字，存在于书本或文章之中。对于景观设计师来说，搭建结构并绿化空间、创造出可以穿行、体验、欣赏的事物是很重要的能力。这是我们能够通过回顾来评判已做的项目、评价自己的想法及其技术、材料的运用效果的唯一途径。这些方面勒华都做到了。他做研究，进行智力创造，并付诸实践，如此一来，也让人们能够对他的景观艺术做出回应。他的成就十分了不起。他是位伟大的艺术家。作为景观设计师，我们必须听取客户的需求，尊重他们突发的奇想。

吴欣：没错，在为人们所用，并引导人们以一种全新的方式亲近自然世界时，景观设计师的作品就会显得更加有意义。结束此次访谈之前，我想问问你的"社会性花园"[①]。我所接触过的杰出景观设计师遍及欧洲、美国和亚洲，但仅有为数不多的几家工作室对社会性工程有过贡献，你的工作室就是其中之一。

董特：这种想法也是源于一次巧合：有一次我在伦敦和一位女士聊天，她大声地质问我："你怎么可以只为富人服务？你被私人项目宠坏了，你对社会没

[①] "社会性工程"（Social Project）是欧洲福利国家的典型说法。这类工程旨在以住房、健康、教育和其他服务的形式将财富重新分配给不够富裕的人群。

有丝毫回报。"我当时呆住了,但她说的一点儿也没错。我总是对公园和花园很感兴趣,却基本上没涉及城市项目。从那之后,我便开始关注自然在人类生活中所扮演的角色,并寻找破败的城区,那里充满了社会难题。我希望自己设计的项目能够切实改善市民的居住环境,而不只是做表面工程,也不是对那些生活在恶劣现实中的人们聊表关慰。我不想干涉弱势群体抗议其恶劣生活条件的自由,而是尊重他们的个体特征。社会性项目是对这一人群的关怀。我想让他们感受到,这样的工程是真真实实为他们而建的,是能够改善他们的居住环境并保留他们独有的特点的。每隔一段时间,我的工作室就会在某片贫穷的城区兴建这样的工程,这些地方经常居住着外来移民家庭。这些项目毫无利润可言,我们完全是以服务社会的宗旨来做这些项目的。

吴欣:这项工作是如何开展的?你会先选好地址,然后再请求市政府批准吗?

董特:不是,这些项目都需要进行招投标。从逻辑上讲,对于这种非重要型工程,市政府希望建设方案的成本越低越好。因此我们总是给出很低的报价。这样低成本的工程从建筑材料到施工过程都需要社会的参与。这类工程我们已经在布鲁塞尔附近修建了3处。效果很明显。人们惊讶地发现这些景观并不会随时间推移而残破不堪,而且还能让居民在房子附近就能够亲近到大自然。有钱人为了躲避城市环境的压力而纷纷逃离都市生活,前往乡下居住。穷人缺乏交通工具,只能依赖住所附近的自然设施。我们逐渐意识到,即使是很小的改造也能带来巨大的变化,并能为场地带来美与爱。这是景观设计所能为现代社会做的贡献。

吴欣:希望中国的景观设计师能从你的理念中获得启发。谢谢!

(蔡金栋 译,田乐、吴欣 校)

Erik DHONT
—between contemporary aesthetics and Europe an garden tradition

Erik DHONT is a European designer with an office in Brussels. A graphic designer turned into landscape architect, Dhont is known for his integration of abstract forms, local biotope and European garden traditions. It is striking that his creations as a landscape architect comprise so many gardens, and that, as a contemporary artist, he is engaged in historic garden restoration. Approaching landscape architecture as an art that engages, in a creative manner, with both a shared European tradition and a contemporary aesthetics, Dhont has successfully bridged the historical and the present, the vernacular and the universal. To him, the old tradition and the new aesthetics should, and can, both be embraced, because each of them rests upon an understanding of culture as a historical process of re-interpretation of the past and of exploration of untapped possibilities of human action. Dhont's broad interest ranges from private garden, to historical landscape, to public environment, to horticulture, to graphic arts, furniture design and video installation. Designing for "a dream of the future", his works have addressed the needs of clients from business tycoons, to art collectors, to young professionals; from historically conscious philanthropists, to fashionable urban chic, to disadvantaged immigrant communities. Dhont mainly works in Belgium, the Netherlands, and France. In the recent decade, he has been invited to design for a diversity of international projects. He is also a frequent lecturer at design schools, and the subject of a multilingual publication in English, French and Dutch — *DHONT: Gardens, Hidden Landscapes* (Ludion, 2001).

Finally, in relation to the theme of this issue, I would like to locate Erik Dhont's effort in the cultural and artistic context of Europe. Garden aesthetics was the center of many aesthetic debates in 18th century Europe. It was then moved out of the spot light in the 19th century following the rise of eclecticism and imitation of historical styles both of which discouraged artists from creating new garden art forms. At the time of 20th century development of modern art (impressionism, fauvism and cubism, etc.), there was virtually no development of a modern garden art in Europe, and this lasted until WWII. So by the middle of the 20th century there was no longer any aesthetic tradition to which garden designers and their patrons could refer. It is only in the postwar era, a few insightful contemporary artists, such as Bernard Lassus in France, Ian Hamilton Finlay in Scotland, started to reconnect art and garden. Erik Dhont has learnt from the Situationist International, an important movement in the 1960s at a period of great intellectual turmoil that saw the stirrings of the Green Movement in the Netherlands, Scandinavia and Germany. In particular, he has learnt from the experimental art of Dutch artist Louis Guillaume Le Roy, a follower of the Situationist International. Le Roy gained international fame as an advocate of new attitudes against unlimited consumption and destruction of nature. He advocated a new balance between spontaneous nature development, and everyday human interaction with nature (with the exception of mechanical or chemical interventions); and called for appreciation of the self-organizing capacity of nature as a source of aesthetic enjoyment of its human users. Obviously, Le Roy's perspective was completely different from the one proposed by Ian McHarg in America at almost the same time. Erik Dhont has transformed the aesthetics proposed by Le Roy, by calling for attention to the balance between artistic intervention of humans and the self organization of nature. This is a far more complex goal, opening lines of an artistic creation in harmony with natural environment. Modernism had introduced a manicheism aesthetics in the arts, opposing without any possible compromise modernity and history. Our contemporary world has questioned this manicheism in almost all domains of artistic activity. Yet, landscape architecture seems to be still under the spell of the call for the tabula rasa, the clean slate where modernist designs free from the ferrets of tradition could be erected. The works of Erik Dhont demonstrate that contemporary landscape design can move beyond that stage and reinvent itself. To this end, his works contribute to a renewal of garden art in the West.

Xin WU(WU hereafter): I was very impressed by your built works when visiting Brussels several years ago. Today, I would like to focus on the thinking process and method behind, in particular your design philosophy. You have been very successful in introducing contemporary designs in historical gardens or landscapes. I propose that we take this as a starting point, since this is a very important issue in China, where the confrontation of historical landmarks and contemporary developments is somewhat comparable to the situation in Europe.

Erik DHONT (DHONT hereafter): As a student, I was already very interested in contemporary art and also in history and historical transformations of culture. Landscape architects who engage in the restoration of historic gardens, at present, place much emphasis on heritage restoration. They strive to reconstruct the formal appearance of historical gardens when they were in their prime, even though they are usually not really successful in this respect. One may remark, however, that we no longer live and think like the people of four hundred years ago in the 17th century, even if their gardens are revered as part of the national heritage in France, England, Belgium and the Netherlands, to name just a few countries. So there is an unbridgeable gap between the way people lived in and interpreted the environment in the distant past and at present. The past is a foreign country and speaks a foreign language that requires translation in order to be understood today. This applies as well to gardens and landscapes.

In landscape architecture, nevertheless, the logics of the natural world and of human scale and proportions are unchanged. This means that you can make a historical garden into a contemporary place provided you maintain a balance with the environment that it had established. You have to respect the measurements, the materials and the scale of the place. You can still choose now the traditional planting materials of the historical period to restore the garden, even though the design can be utterly different. Each historical garden is a repository of historic values, and each one is unique. There is no universal recipe to follow. Yet one can agree that it does not make sense to change long established values. As a designer, you can reuse these values and still make a contribution to contemporary art. Very often at present, designs by landscape architects who want to assert their contribution to contemporary art are too aggressive, may be too eager to make a show of the ego of the designer. I think one can be more modest and more respectful of the environment. One may successfully restore historical gardens and translate their values into contemporary culture if only one is aware of nature in the microcosms, and also in the macrocosms. These are the two fundamental values of our time for me as a landscape architect.

It is also important to think of the maintenance, of the management of the garden after it has been restored. You can maintain a garden perfectly in the shape in which it was originally build, but it is somewhat unrealistic to plant trees and put much effort in preventing their growth. It seems to me more important to establish a good balance in coordination with nature. For instance, you may design fifty percent of the project and leave the other fifty percent to nature. In such simple fashion you introduce a good balance between humans, plants and wildlife. And the wilder you leave it, the more birds will visit and dwell there. I am always very conscious about my position as a designer. Planting a tree is going to change the landscape that belongs to many, we must feel responsible for its future and well-being.

WU: This reminds me of the project at Gaasbeek. Your contemporary design was fortunately supported by Herman van den Bossche, a heritage researcher at Flemish Heritage Institute. But in most situations, many people think that the historical and the contemporary do not go together. Do you think there is an issue of aesthetics?

DHONT: Yes, that is a problem. Aesthetic values keep changing, reflecting artistic innovations. I talk about design, that is, about aesthetic values. Many people, including many designers, believe that innovation is exclusive of tradition, but I think it is possible to maintain the tradition of garden design, while introducing the aesthetic values promoted by the environmental and ecological movement. In such a way, we can both enjoy references to garden tradition and have the pleasure of looking to a garden in balance with nature. The beauty of tradition can be translated into contemporary design. For example in 16th or 17th century gardens you had pyramids, spheres, animals and characters of clipped box that were used to produce classical topiaries. We can also do such topiary[①] in a contemporary way, giving them a more abstract form, since we do not like as much figurative as abstract images. So you can produce an interpretation

---

① "Topiary" is a modern word derived from an ancient Latin word, topiarius, the gardener. This word itself was introduced in the Latin vocabulary when gardens became fashionable in Rome, at the very beginning of the Roman Empire, just before the beginning of the Common Era. It was borrowed from the Greek, topia, (meaning literally: places), which designated in Rome the different places that composed a landscape in wall painting. So the Roman gardener was compared to the painter who creates the illusion of real places with paint.
However, in modern English, topiary was used instead to describe the way in which gardeners were creating geometrical forms or figures (such as a single animal or character, or even an army or a group of hunters pursuing wild animals). Topiary had been very fashion-

of tradition that leads to a different design. You may also call attention to the reflection of the light on the plants, or things like that. However, in contemporary landscape architecture, it is also important that the owner of a historical place understand the possibility of keeping these values. You do not need to always do some thing new; you can first maintain what you already have, and then you can put some contemporary accent, and that may be enough.

WU: You said "translation" . Can you explain in a real project, for example the project in Gaasbeek, how and what did you translate?

DHONT: That place dates to 1602. The garden has been built now. This was a historical place where there was a Renaissance garden. It means that you have box hedges, flowers and a beautiful garden in rectangular beds. In the Renaissance, a labyrinth was a very important feature to make a garden pleasurable, to make it a "plaisaunce" as a garden was called. So we proposed to introduce a labyrinth as a translation of the place into a Renaissance garden. The translation is historically correct, but the design had to be contemporary in order to make sense of a walk in a labyrinth according to contemporary ideas. So the design does not follow the pattern of a classical labyrinth, but rather proposes an abstract labyrinth, where you have some design details or scent like the roses which lead you until you loose yourself in nature. So that was the abstract idea for the garden that we have built. It was built in 1992 and now, after 20 years, you can see that the garden is developing very well in time. It is important that plants themselves and the trees grow well, and flourish in the future. As a landscape architect you imagine what a place should become, but you do not

---

able in Northern Europe in the 16th century, but the elite turned down their nose on this practice during the Baroque age, and it is only with eclecticism in the 19th and 20th century that an interest for old topiaries was revived. Some of the most spectacular topiaries have been created in the late 19th and early 20th century in England, in particular at "the Sermon on the Mount" at Packwood House, where spectacular yew trees are clipped into an abstract rendering of the crowd gathered to listen to Christ (ca 30 CE). As we shall see further down in this interview Erik Dhont is a rare contemporary garden artist who has tried to rejuvenate the art of topiary while availing himself of the resilience of the same Taxus baccata, the common yew, in use at Packwood.

know whether it will really happen, and it is a pleasure to come back in a place after twenty years, and still feel the power of the design and the beauty of nature and of its development. One has to be careful, because when you use too much contemporary material, or too much material that degrades quickly with time, it may ruin the whole project in the future. This is why I like to use natural stones, rather than concrete, because it ages much better. Of course there are modern materials that you ought to use for structural reasons.

WU: When speaking about the design of a labyrinth, you have mentioned — "what is the idea of a labyrinth, is it not an entertaining route that leads you to confront yourself?" Is this what you meant the abstract idea of a garden?

DHONT: Yes. There is a constant effort in all the gardens that we have been designing. It is our struggle against the separation of the house from nature. Dwellers experience a deep psychological attachment to their house, but our main purpose is to get them to distance themselves from it and to engage with nature. A garden provokes the curiosity to take notice of nature, to reach a view of the environment, to fall in love with it. It is the garden that will bring you to observe things in nature, but also to be calm, to be detached. It is much more important than before. Recently we have designed a secret garden for a place in Holland. It is about loosing yourself in the garden, entering a dream and being happy with plants around you, and more and more we are aware of the aesthetics, and the pleasure this may yield.

WU: I always find a conversation with you delightful. When talking about garden, you use words such as love, respect, joy, making a dream...concepts that help one to understand gardens are places with many layers of emotion, beyond architectural space.

DHONT: The nature of architecture concerns the content and function of a place. As an architect you must make structure, endow places with structure. I think that landscape architects are dealing with something else. Landscape architecture does not address a necessity on the primary level. It does at a secondary level. It concerns looking, smelling and appreciating. That secondary level, in fact, reaches for a higher awareness of the environment. The content that a landscape architect is dealing with partakes of the nature of a dream. We are called for making a plan and proposing a dream, but of course we plant the plan and the plants will grow and the dream will come true through time. Many people ask a garden, but they are impatient, and as a result the most beautiful gardens on design drawings are dreadful in situ because they tried to ignore the evolution of time. It means that we ought more and more heed the fact that time influences the beauty of a place. At the beginning, a design is always a way of proposing a dream, or a way of living, or a way of looking. But the patrons do not look with the same gaze as a designer does; we invite them to look differently on their environment, sometimes restricted, sometimes open…it depends on the location of course.

WU: What a beautiful metaphor of a garden as a dream to be fulfilled! So the development of the garden as time elapses is given much attention in your design. Even when you plant your trees you keep thinking how the trees will grow and how the environment will respond to their growth. This extends the responsibility of the landscape designer, does not it?

DHONT: Of course. It is very important to look at old gardens, because there you can see how things have developed in time. Historical gardens give you a walk through the ways in which people were behaving with respect to the environment in the past. There you see how trees have thrived and how their impact on the environment. We are really learning from past generations. Starting from that knowledge, we can proceed with the translation to contemporary society, knowing that, at present, we live differently. However for humankind, the enjoyment of beauty is still the same, so we see that people are still looking forward to the happiness beauty can bring to their lives.

WU: When looking at a historical place, you pay attention to the place not only as cultural heritage, but also as natural heritage.

DHONT: Definitely. Let me insist on this relationship between cultural and natural heritage. Very often you look at a garden for the main axis [①]. So often when

---

① Since the Italian Renaissance, great pleasure gardens in Europe have been organized, as much as possible given the constraints of their location, in symmetrical fashion around an axis perpendicular to the facade of the main house, the castle or palace of the lord living in the place. This axis received more and more attention and, starting from Vaux le Vicomte and Versailles, the gardens of Europe have followed the fashion created by Le Notre, making this axis into a line running to infinity. This is the axis to which Erik Dhont is alluding here.

you look at a garden you see the main direction, but that direction also gives you information about power, the power to build and the power to show. Yet, the most interesting axis to me is the perpendicular axis, because there you have the romance of garden, the other way to look at the place. Louis XIV had written a little text about the ways of showing the gardens at Versailles [1]. It was about the ways of looking at the garden, the fountains and the palace in order to derive the most pleasure. It was about the pleasure of looking in such a way that you have the enjoyment of one view, then of another, another, and so on. It was just a way to make visitors aware of each place, because every place has to demonstrate its own logic without any explanation. This is what is beautiful about landscape design or gardens, it follows a natural logic. When one designs with natural logic, the work achieves much better results and also greater beauty.

WU: I remember you talked to my students about how to look at Bomarzo [2]. It was most inspiring. Let us now turn to your design approach. First, how did you move from graphic design to landscape architecture?

DHONT: It was a coincidence. I spoke to a friend who had a nursery of fruit trees, ornamental plants and flowers. He noticed my interest for garden and landscape design, and opened my eyes to that. The graphic school was more a school of typography, design of books and letters, which has helped me to look differently at design and pay much attention to proportion. The world of plants is a world that you

---

[1] There are several pages written by Louis XIV himself to explain how the gardens of Versailles should be shown to foreign ambassadors which are still extant, and have been studied and published by historians. See Simone Hoog, *Louis XIV: A Way to Show the Gardens of Versailles*. And more recently: Thomas F. Hedin and Robert W. Berger, *Diplomatic Tours in the Gardens of Versailles under Louis XIV* (Philadelphia, 2008).

[2] The Park of Bomarzo was created, in 1552, by Vicino Orsini, related through his wife to the Farnese family. The famous Italian archaeologist of the 16th century, Pirro Ligorio, is said to have designed the garden, with its fantastic sculptures. It looks like no other garden of the Italian Renaissance, and went into oblivion during the 19th century. Until it was given much publicity by surrealist poets and artists in the first half of the 20th century. Its fame has never ceased growing ever since, and it has inspired a large number of contradictory interpretations as well as many art works.

study and from which you learn. I am still learning everyday, because any environment is different.

WU: Many people are fascinated by your sketches of garden ideas. Can you tell me about their role in your design work?

DHONT: Sketching is a way of thinking about things, of exploring how space can be defined or how action can be located into a plan. It is a way to imagine a walk through space, or to imagine a solution to a problem. Any architect or landscape architect runs the risk of being too easily satisfied with concrete necessities. I think that in landscape design, after you have studied the material constraints very carefully, sketching gives the possibility to think more freely, to take distance from all these constraints although you have them present in the head. The sketch is a way of liberating oneself from the demands of technique and function, and of discovering other less materialistic dimensions of a garden experience. Later the possibilities opened by the sketch would have to be confronted to the existing constraints to which the project should bring a solution.

The next phase is finding a balance between the idea in the sketch and the constraints of the practical world. The sketches are really meant to take distances from my thinking habits in everyday practice or for training myself without any project in mind. Just to give an example, when you sit on a plane, you can make sketches of ideas. Far from idle doodling, these sketches are very important to me, precisely because the ideas, which they explore, are not related to a project, but to helping develop my capacity of free thinking about plants or space. This is very important for me personally; it is a form of lateral thinking [1] that needs to be developed. In practical terms a designer's active life should be shared between free thinking and engagement

---

[1] Lateral thinking is a term coined by Edward de Bono in his first book on problem solving *New Think: The Use of Lateral Thinking* published in 1967. This book was followed by a large number of other simple books, which all suggest ways of developing a capacity for creative thinking through an indirect approach. They provide a popular version of a large domain of scholarly discussions about the role of induction in creative thinking. Here Erik Dhont is not repeating a method he would have learnt in a book on creative thinking, but rather expressing his own approach.

with real world constraints and problems. To achieve progress in your own work, you ought to develop yourself, and producing sketches offers a way for achieving this internal evolution, for increasing your metaphorical stock. Then, when you are confronted with problems in the design of a garden, you can draw upon this self developed metaphorical stock to propose an innovative design solution.

WU: Your remarks would certainly ring a bell to today's computer-oriented generation of young designers. Many of your sketches remind me of abstract painting. Is translation from abstract painting to landscape design the mode of development that you are pursuing through the sketches?

DHONT: I think it is rather looking at landscape architecture as an art form. One is always a child of the environment in which you live; when you look at new contemporary architecture or paintings it inspires you new ways of thinking about the accustomed world you live in. This personal translation of another artist's work into a new approach to issues in the contemporary world is interesting; but it is not yet art, it is a way of looking. Artists are concerned about creating artworks. For landscape architects it is important to see artworks, because it enables you to see how artists are taking position about the present world and about aesthetics. It enables you to think freely, away from professional concerns. In this way, it is inspiring. However, we should not forget that each art has its own medium. The art of garden and the profession of landscape architecture are about spaces in nature. This has been true since the ancient time.

WU: Is it correct to say that your landscape design is an abstract art about space in nature?

DHONT: Yes, I think it is very precise. This is the purpose I am pursuing.

WU: When you are drafting plans for a garden, the tension between the different parts are essential in its graphic composition. How do you translate that tension into built space?

DHONT: You can translate the tension when locating plants in the garden space, until you reach a balance. When you work too much, or you attempt to bring too much content into the space, the tension will fall apart. It is very important to stop at the right time, to bring energy in a place, but also to know when to stop; because too much content is counter-productive. It is all about balance. One may explain it with many words, but it is about a certain kind of situation, it can be the age of the trees related to the space, it can be the cultural heritage, it can also be the effect of the environment... For example, the valley landscape will be very important to a garden in a valley, because it will impact upon people's feelings in the garden. Tension also results from changes of season and weather, such as the tension due to the light, the wind, the rain, the presence of flowers and birds. These are things that we keep learning out of observation and practice.

WU: How about sculptural form? What place has it in your design?

DHONT: You can sculpt the space of a garden, in the way Isamu Noguchi did in "California Scenario" [1]. There is open space, shadows, and volumes as in any landscape, but on the other hand it is a monument. You can also integrate new sculptural landscaping in a place, in the way Robert Smithson did with the Spiral Jetty [2]. It modified our gaze upon the landscape. Land artists have made many interventions of that kind in the 60s and 70s. That was 40 to 50 years ago, and now is

---

[1] "California Scenario" is a public garden, commissioned in 1979 by developer and philanthropist Henry T. Segerstrom, in Costa Mesa, California. Segerstrom, who asked Noguchi to design a public garden to enhance two office towers built on family land once used as a lima bean farm wanted a lush retreat. Noguchi instead created a simple stone plaza with a few green spaces. The Noguchi Museum writes that the artist first conceived the project as an "abstract metaphor for the state of California, from the Sierras, to the desert, to the woods. In addition to including redwoods and cacti, among other native plants, it encompasses a number of individual elements designed by the artist to evoke some of California's salient characteristics." It is quoted from the announcement for an exhibition in 2010 of pictures of California Scenario by the Noguchi Museum in New York City.

[2] The Spiral Jetty is certainly the most famous work by Robert Smithson. It is an earthwork constructed in a desert shore of the Great Salt Lake in Utah. It demonstrates the possibility of creating artworks with landfill, a not so noble material, in a region where there are no visitors, to the difference of a Metropolitan Museum. In both respects this was a manifesto for a renewal of the New York artworld.

a different time. Their creation was very important then. Now it is inspiring for us to look back and to relate land art to the greater landscape tradition of the west.

WU: How do you treat sculpture directly in a garden?

DHONT: Recently, I have just finished a new garden in the US, featuring the contemporary sculptural collection of the client. It raises a specific issue, since you do not want the different sculptures to interfere with one another. It calls for modifying the landscape so that we hide what we do not like to see, thus affording you another view of the environment. When making this sculpture garden, I created different zones so that each sculpture has its own visual integrity. This is the way in which you can sculpt the landscape to trigger an idea, or to stimulate a sense of intimacy, or to present a different view of the environment. These are different ways in which landscaping and sculpture can meet and dialogue. When you have less space you can achieve a similar effect with hedges, or topiary, giving them the shape of rocks for instance. I use yew trees (*Taxus baccata*). It is both natural as a tree, and manmade as a clipped artificial form.

WU: I would like you to talk about how learning from your own ideas. Most designers today do not, even avoid to, go back to see completed projects. You revisit your old projects, and your team keeps up with their maintenance. You even establish a work mood called the "three variations" in order to explore an idea with different media. Can you tell us more about this?

DHONT: When you look back to your own design, you look back to your own evolution. When going back to an old project you can see what has worked well, and what was a mistake, and you learn from your own experiences. When you had an idea or a solution for one garden, it was for a specific place, cannot be copied to another case. However, you can make a variation on the idea to explore its possibilities. Yet when you make too many variations you start copying yourself and your creativity is no longer active. That is where my concept of the "three variations" comes from.

For designers, it is beneficial to rethink their work, to learn from themselves and to develop a few variations on some brilliant ideas; but it is most important to limit the quotation of one's work and to always evolve with time. This constant progress compounded with variations helps you develop a capability of inventing new solutions to different design problems.

WU: Do you mean that, like your practice of sketching, this is a means to the development of a personal design style?

DHONT: I do not aim at a style, but a signature. That is, a personal way of developing designs solutions, a way of doing research and of analyzing problems. This is what I call a signature. Style speaks for something else. Style is related to fashion, it is a consolidation, a freeze imposed upon potential for invention. To the contrary, a designer should always remain free, and push oneself hard, to explore untapped possibilities.

WU: So the freedom of innovative thinking is something very important for you as a designer.

DHONT: Absolutely. Nevertheless, in real projects, there are always practical issues and constraints that put different kinds of pressure, which is also good.

WU: You have worked in cooperation with artists using different media, such as modern music and dance. You have also designed your own line of furniture.

DHONT: I think about the dance. The beauty of the garden provoked people into dancing. It was just a spontaneous happening. The furniture is more a part of design: how can you make a bench that will last for ten years, or a chest that will last for thirty years even if you leave it outside? I researched on the great crafts traditions of Belgium. When you look at the different contents from music, furniture, design, architecture, everything is greeting to each other. It is a way of being aware and exploring.

WU: When landscape architects mentioning research, it is usually about finding useful information from studies done by scientists, such as ecologists, botanists. It is unlike your kind of research, which is carried out by yourself for the sake of improving your design ability. What are the main topics of your investigations at the present?

DHONT: At the moment I am investigating the tradition of garden forms, the tradition of topiary in particular. What are the possible forms of topiary? It can be a cube, a prism, a pyramid, or any geometrical form, but it can also be an organic or a plastic form. I have pursued this exploration in different projects, and I must say it could be extremely sculptural and also very nice to do. I have also been researching how many steps we need before we look right, and before we look left, when taking a walk; and also, how to do things and not to do things, when designing a garden. It is important to design to give content to a place, maybe to add a last touch to the surrounding space and also to focus the eye, to highlight the experience on the way of looking to plants or to nature, but to reinforce this content may be counterproductive, may override the balance between human intervention and the free play of nature. It is a part of the job of landscape architect to maintain this subtle balance.

WU: Since you mention the dilemma of acting or not acting in a landscape, it reminds me of your admiration for the work of Dutch artist Louis Guillaume Le Roy,

---

① See also *Nature Culture Fusion-Louis G. LeRoy's Eco-cathedral*, quotations by Louis G. Leroy compiled by Esther Boukema and Guido Van Overbeek, Photography by Philippe Velez McIntyre additional texts by Piet Vollaard, Hagen Rosenheinrich and Vincent Van Rossem, in English and Dutch, published by the Netherlands Architecture Institute, now available from Boekie Woekie books by arstists Amsterdam, www.boekiewoekie.com
Louis Guillaume Le Roy is a Dutch descendant of a French Hughenot family. He was born in 1924 and is still living in the Orangewood neighborhood of the city of Heerenween in the Netherlands. His early works in the 1960s and 1970s, which were critical of the lack of attention to nature and her capacity for self organization, have drawn much attention in Germany and the Netherlands. After he engaged in the transformation of a one kilometer long and 15 meters wide separating ground between the double lane streets forming the Kennedylaan, in a residential neighborhood, he received many commissions in Germany, which all failed , because they called upon interactions between humans and nature, which displeased the "deep ecologists", and also because he insisted on the free play of both nature and humans, which displeased local administrators who believed they had a duty to impose some rules and limits to anything within their reach of action. His thinking was very much inspired by two French philosophers, Henri Bergson (1859, 1941) and the much younger Guy Debord (1931, 1994) as well as by the artist Constant Nieuwenhuys (1920, 2005) and the Situationist International which Constant Nieuwenhuys founded in 1957. A

in particular his "Eco-Cathedral" in a nearby Dutch town Mildam [1].

DHONT: Le Roy is an artist who has given much attention to the postwar ecological movement. The ecological movement, however, was not very articulate, to say the least, about aesthetics. It has been able to explain how we could help certain vegetation to survive and maintain itself in a landscape, but it has been very weak on creative design. Le Roy's book — *Switching off Nature, Switching in Nature* (Natuur uitschakelen, natuur inschakelen), 1973 — struck me greatly. It was the first time that I heard of re-using waste material for landscaping purposes, re-using bricks and stones from demolition sites to build new kinds of gardens instead of throwing them into dumps that would further contribute to environmental pollution. His life's work, the Eco-Cathedral, was about putting these recycled materials together to achieve a new monumentality. More importantly, he explained how he would do things, and also avoid doing things; how he would balance action and non-action. On the one hand he would build a supporting landscape structure with waste materials allowing everyday use by neighbors to carve paths and simple amenities for by-passers; on the other hand, he would let nature overgrow most of this supporting structure and develop on its own. He creates a balance between his action and his respect to nature. What is crucial, however, is that it is an interaction that opens the way to aesthetic appreciation of a landscape form which arose from the free interactions of humans and nature. This

---

grandchild of Louis Guillaume LeRoy, recently wrote on his website: "My grandfather is an artist, philosopher, writer, painter and photographer." He is quite right, and in fact the writings of the Nobel prize of biochemistry Ilya Prigogine on the production of order by organic nature, has proven very important for the development of Louis Guillaume LeRoy thinking. He started a personal project on a tract of agricultural land he had purchased, about 30 years ago, the Eco-Cathedral in Mildam. All the building materials (rocks and stones) from the city Heerenveen were brought to his land and he started to build a "cathedral" from it. Shortly: his philosophy is that one has to bring nature in contact with the rubble we do not longer use, and that we can introduce structures which are expressive of our own way of life, with our hands, and let nature organize life around this work. This will allow nature to demonstrate its capacity for self-organization, and will reveal structures which are beyond the scope of human planning. In the discovery of these structures lies the beauty of such landscapes. Single-handed he has built all kind of sculptures, paths, stairs etc. on his land. Nature has captured them, grown in and over them. He vows that this project of interaction between man and nature should continue until the year 3000.

appeared to me as a way of paying respect to nature without relinquishing the role of design.

WU: I notice you used waste materials in several of your designs. Even in a project for a trendy young couple, you were able to convince them to accept such a challenging new aesthetics.

DHONT: In general, my clients are extremely cultivated. But you ought to have ideas in order to convince people. Designers have to be able to convey their ideas; when they fail their ideas remain in a written form, in books or articles. As a landscape architect, it is very important to build and plant places, to do things that can be walked through, experienced and appreciated. It is the only way in which you can judge, in retrospect, what you have done, the only way that you can evaluate your ideas and their technical and material implementation. Le Roy did all of that. He did the research, the intellectual invention, the artistic work and the practical implementation, thus allowing people to respond to his landscape art. His achievement is remarkable. He is a great artist. As landscape architects, we are subjected to the demands and the whims of our clients.

WU: True, the work of landscape architects is significant when it is used by people and leads them into a new form of engagement with the natural world. Before finishing this interview, let me ask you about your "social projects" [1]. In my own contacts with the best landscape designers in Europe, the US and Asia, your office is one among the very few who contributed to social projects.

DHONT: The idea resulted from a coincidence: I had a talk with a lady in London, and she exclaimed "How dare you work only for the rich people? You are spoiled by private projects; you are not giving anything back to the society". I was taken aback, but she was right. I had always been attracted to parks and gardens, not to urban issues at large. So I started to pay attention to the role of nature to humans, and looked for decaying urban areas where there were social problems. I wanted to

---

[1] "Social project" is a typical phrase in the European welfare states. It means that these are projects which aim at redistributing wealth, in the form of housing, health, education or other services, to the less wealthy tier of the population.

design projects that would alleviate residents living conditions, not cosmetic projects that merely pacify the spirit of people living in harsh realities. I want not to interfere with disadvantaged community's freedom to protest against their situation, but to show respect to their personal character. The social projects are a compliment to these people; I wanted them to feel that this was really a project meant for them, that they could contribute to the improvement of their own environment and its unique characteristic. Once a while, my office would develop such a project in one of the poor neighborhoods, many inhabited by immigrant families. No profit for the office, we really do these projects with social aims in mind.

WU: How was it arranged? You look for the site and then petition for approval by the city?

DHONT: No, there is always a bidding process. Logically, the municipality looks for a proposal that is the least costly for insignificant project. So, we bid in a very low price. These are low-cost projects that call for a social form of participation, from materials to construction. We have made three such projects around Brussels. It is working. People are astonished to see how these landscapes are surviving over time, and how much it allows the residents to relate to plants in their neighborhood. Wealthy people evade the city to go to the countryside to rest from the stress of urban environment. Poor people who do not have the means to travel depend much more on the natural amenities close to home in their own neighborhood. We have come to realize that even very little transformation can achieve great changes, bring beauty and love to a place. This is the kind of contribution landscape architecture can bring to contemporary society.

WU: I hope that Chinese landscape designers will be inspired by your ideas. Thank you! (Translated by Jindong CAI, Proofread by Tina TIAN, Xin WU)

林璎
——艺术创作与形式

大多数人是从林璎年仅21岁时建成的第一个作品——位于华盛顿国家广场的越战纪念碑了解她的。越战纪念碑吸引了美国公众注意,并引发许多争议,同时也第一次使林璎有意识地感觉到了她的华裔身份。在1982年5月纪念碑落成后,她于1985年与父母一起首次回中国大陆,此后,她经常回中国。

艺术家要摆脱成名对他们的影响,需要勇气和决心,很多人一直做不到这一点。林璎花了将近10年的时间,才逐渐摆脱越战纪念碑的光环,并在纪录片《林璎:一个清晰而有力的视像》中谈论纪念碑的设计。同时,她拒绝将自己定格为纪念碑设计师,在过去28年中,从未停止过对艺术的追求,从工作坊艺术作品到户外雕塑和装置,到花园和公园和建筑,创作了丰富的作品。2006年,林璎出版了她的第一本著作《交界线》,在书中她写道,"我感觉我存在于科学与艺术、艺术与建筑、公共与隐私、东方与西方之间的交界线上。我一直努力在寻找不同力量之间的平衡,寻找对立面相遇的地方。"

今天我遇见的林璎,是一位自信、率真、成熟的女艺术家,积极投身于与全球环境和中国文化相关的活动。但她认为自己的策略是非政治性和非说教性的,她相信历史有成就未来的力量。就在本次访谈的几天前,林璎参加了2009年联合国哥本哈根气候大会,并在会上演示了她最新的视频作品"再生一棵树"。该视频是她目前正在创作中的一个作品"什么正在消失？"的一部分。这也将是她设计的最后一个纪念碑项目，关注地球上已经消失或者即将在我们有生之年消失的物种和场所。林璎向世界传达了这样的信息：我们每一个人都能对环境的改进有所贡献。2009年9月，林璎帮助启动了纽约中国城的美国华人博物馆，她为博物馆进行建筑设计，并担任董事会成员。在最近二十年中，她与大提琴家马友友和现任美国商务部部长、原华盛顿州州长骆家辉等华人精英一起积极发动海外华人社会。林璎经常访问中国大陆，她认为中国设计师应该珍惜中国悠久的文化和历史，在新与旧之间寻找平衡。

吴欣：你总是称自己为一位艺术家和建筑师，景观在你的艺术和建筑之间扮演什么样的角色？

林璎：景观是我的艺术和建筑的一个有机部分，而且经常是我的作品的主题。在我的建筑中，景观与建筑造型不能分隔开来。我的建筑不是仅仅立在景观中，而是积极地与环境互动——常常是通过营造与室内空间无间交融的室外空间；这样的建筑同时创造出一个体验和欣赏自然的框架。在我的艺术中，环境、地形和区域是所有雕塑作品的灵感和文脉。

吴欣：所以景观是你的艺术和建筑之间的连接？

林璎：是——但不仅仅是在形式上——我的建筑注重尊重和结合自然环境的可持续性理念。我想我的有些艺术项目设计意味较强，或许可以叫做景观设计。然而，像"读花园"和"输入"等花园项目也融入了语言，是以概念为基础的。这些作品和纪念碑一样将语言和信息蕴藏于环境景观之中。在名叫"食"的大型城市公园项目中，我使用了时间信息——在中心溜冰场下是一个用光纤构成的 2000 年除夕夜该出现于该城市上空的星相。这个作品标记了时间上一个特定的点。该公园是为千禧年设计的，滑冰场上的星相图记录了一个重要的时刻。它在文字信息层上添加了科学性的数据以组织空间。

吴欣：说到文字，你是运用它来加码空间？

林璎：是的，语言是一种物质材料。它不一定必须与物质形体相重叠；实际上有时候它就是形式本身。人们可以说越战纪念碑是一个纯粹的表面，并不是一个凝重的物质实体；然而，那些刻于抛光的土层平面上的名字是实体的存在。字体的尺度变得至关重要。我们通常习惯于阅读纪念碑式的大字，却很少使用近人的书本大小的字体——这样一来就自然地要求一种更个人化的阅读，在心理上变得更亲切和感性。

我们有 5 种基本感官——视觉、听觉、触觉、嗅觉和味觉；而我认为人还有另外一种感觉官能，那就是阅读。阅读行为是人类的发明，没有任何其他物种有阅读的能力。我将阅读视为第六感。我们不是看文字，而是感觉文字。我非常着迷于在我的作品中使用这种感觉。

吴欣：你说的阅读，是指从离散的信息中构造意义的能力么？

林璎：阅读是一种非常复杂的物理和情感行为。通过以非常综合的方法在这些作品中展示阅读，我寻求一种人与人的沟通。它不是孤立的——我创作形式，然后传达信息——将两者完全结合。当你制作一个雕塑时，它是完全沉默的；人们只会对它做出相关的反应。然而，如果你说"读我"；即使观者原来只打算摸一下它，将会出现什么情况呢？我无法悉知观众们的反应；我所可以肯定的是，阅读为艺术作品增加了一层含义，亲密感和人性。

吴欣：是一种开放性的"输入"？

林璎：当我和我的哥哥林谭合作"输入"这个作品时，文字和形式是在同时生成的。他创作有关记忆和时间的诗篇，以及记忆是如何根据随意而非线性连接来进行工作。而我则设计了花园。人们去参观"输入"花园，问要怎样才能"品读"这个花园？是否有必须遵循的路径？我说：没有，这就像回忆，你通常是在无意中找到回忆，然后便与过去的事情连接起来。它是我与哥哥关于阿森斯市俄亥俄大学校园的回忆，那是我们成长的地方。然而，一个场所的回忆能够与大家一起分享，任何在俄亥俄大学校园度过一些时光的人们，都可以与他们各自的经历联系起来。实际上我哥哥写了一首诗，我则设计了造型。然后我开始对每个元素进行调整，使其与哥哥的文字相匹配。文字及其意义影响了空间的布局，这是非常有启发意义的。

吴欣：那"读花园"这个项目呢？

林璎：那是我们首次合作，过程很相似。他写了诗；我将诗分为三部分，对应地设计了三条进入花园的路径。他又返回去调整诗句……与物化的路径相联系。

吴欣：与"输入"相比较，在这个项目中诗的视觉效果更强。

林璎：是的，这个作品更具有趣味性。我们希望它能够具有像"爱丽斯梦游奇境"的效果。虽然它并不是专门为儿童设计，但我们希望文字易读有趣。对我来说，整个花园就是有关文字的方向性和影响力。

吴欣：这个作品的确很有创意。文字激起人们阅读的欲望，但又是以一种完全陌生的方式来与语言接触。

林璎：是的，的确如此。文字和诗歌的具体性有一种吸引力，并正在逐渐呈现出来。在诗歌中，有三部曲。哥哥和我正在等待继"读花园"（俄亥俄州克里夫兰市）、"输入"（俄亥俄州阿森斯市）之后在俄亥俄州的第三次合作机会，以完成我们的三部曲。然后这将以书籍的形式出版，以纪念我们的父母亲和我们家在俄亥俄州的生活。所以，它同时也是由花园、诗歌和书籍组成的三部曲。

吴欣：这个创意非常好。你们想借此赞美父母的养育之恩么？

林璎：是的，这些都是温馨的回忆。我的父母亲都是来自中国的学者——我的父亲是一位陶艺家，母亲是一位诗人。他们教给了哥哥和我很多东西，让我们学会发现自己的潜能。我往艺术方向发展，而我的哥哥往写作方向发展。我们自然而然地追随了父母亲的足迹……

吴欣：中国艺术家历来追求"三绝"，将诗、书、画三者融合在同一框架中。

林璎：非常有趣。孩提时代，和很多在冷战后长大的美籍华人一样，父母没有教哥哥和我学中文，我们也从没有正式接受中国艺术方面的教育。你刚刚说的"三绝"，对我来说是个新鲜事。但想起来，图画、语言和书法对我来说都是非常美妙的；而我总是自然而然地为文字和造型所吸引。

吴欣：你的创作当然是非常不同的，不过在你的作品中，有一种美学观点让我觉得非常亚洲化，即你关注事物间的互相关联性而不是孤立的单一物体。

林璎：确实，我不能够分离事物；我必须把它们看做一个整体。就拿艺术、建筑和纪念碑来说，我把它们看成一个三脚架，不能拿走其中任何一条腿。目前我在艺术创作和纪念碑设计上花的时间比建筑多。要真正地在建筑上达到一定的高度还需要更多的时间来产生一定数量的作品。也许还再要十年，因为我每次只能专心做一个项目。我坚信建筑是一种艺术。

吴欣：那您是否把自己看成是一个景观设计师？

林璎：我非常喜欢在大地上工作——并且创造环境。但是花园设计不像我为斯托金艺术中心创作的"波场"雕塑那样，是非功能性的纯粹形式的艺术。无端地在一件艺术品中加入一条小路，把它改造成一个功能性空间，就会将艺术形式的纯粹性毁掉。我非常喜欢设计花园和景观，但是我内心有点矛盾。我非常着迷于雕塑和地景艺术，但是如果赋予这些艺术品功能性，会毁掉它们的艺术性。所以，尽管我喜欢设计城市景观，最好还是不要在创作中再增加一个新专业领域。我的作品向来就难以定界归类。

吴欣：这是许多跨界人士的共性。你在不同领域创作时，如何把握尺度问题？

林璎：无论是建筑、公园还是雕塑，我的作品尺度一般非常大，但是非常人性化，不是纪念碑式的。摄影从来就不能传达我的设计意图。你必须亲身去体验，才能解读我的作品。我能够想象到别人是如何在我的作品中穿行，触摸它，感受它。这完全是一种心理活动。无论是什么作品，它们都是关于观众对艺术的个体反应。

吴欣：所以在你的作品中，重要的是人的体验感受，而不是视觉效果？

林璎：当然。我常说直到人们在我的作品中穿行、体验，作品才成为真正的作品。读完劳伦斯·魏施勒于1982年写的《"看见"意味着忘记看见的东西的名字：当代艺术家罗伯特·欧文的一生》一书后，我认为罗伯特·欧文是一个伟大的艺术家，尽管我当时还没见过他的作品。在欧文的早期艺术生涯中，他创作了一些奇妙的碟型装置,但是禁止摄影,因为这些作品探索了人类感知的边缘,照片不能让人们体味到其中的艺术。

我不仅仅对跨界作品感兴趣,同时也对感知的极限感兴趣,这就意味着需要投入更多的精力去感知艺术作品。当我终于看到了这些碟型装置,就明白了欧文为何不允许摄影。你永远不可能呈现那种经验。另一位艺术家詹姆斯·特瑞尔也是如此,他的作品只能亲身去体验。这两位艺术家对我的影响非常深刻。我非常着迷于艺术是位于体验和感知边缘上的观点。

吴欣:能否谈一下你在实场上的设计手法?

林璎:我必须非常全面地访问和研究设计的场所。我会多次到现场勘查,了解场所特征,把整个场所牢记在我的脑海中,并通过模型记录下场所条件。然后我把这些资料都放在一边,搜索我需要去学习了解的内容。比如在设计密歇根州安娜堡的"波场"雕塑时,我花了将近一年的时间学习空气动力学,才开始着手设计。我学习科学知识、历史、时间与语言。我也通过与人们谈话来收集资料,但更加依赖于自己的研究学习。然后我回到工作室,专心进行创作。通常我最初的设计跟最后的施工效果是非常相近的。我通过研究,从概念层面上去熟悉场所,以此了解我能够为一个特定场所做些什么。对我来说,设计是一个思考和研究的过程。这个过程可能要6个月甚至1年的时间。在研究期间,我尽量不去设计任何东西,甚至不动笔。当所有的研究完成之后,我把这些资料都放一边,让创意浮现出来。我尽量不带着先入为主的想法去看现场,这非常重要。

吴欣:在你的《交界线》一书中,你提到尽量不要太快进入形式。

林璎:对,不然你就会把所有想要做的事塞进这个先入为主的主意和形式里。对我来说,只有当创作与研究平衡时,设计才能更趋向于一个互动的过程。

吴欣:那你怎样知道什么时候那个完美的形式出现了呢?

林璎:我知道;当一切准备就绪时,就知道了。然后最难的是要明白将要做什么。我是根据场所的特性来进行设计的。每一个新项目，我都要根据场所的特征来寻找自己的心声。这是非常艰难的。我常常担心我可能难以将我所想的付诸实践。我不接受有最后期限的项目，因为那样的话，我的成果来自于时间的限制而不是一个寻找创意的自然过程的结果。在我的作品中，形式首先是创意的物质化。

吴欣：在做研究时，你会侧重某些方面么？

林璎：我从没有固定方向的搜索开始，寻求各类有关于场所的物质和文脉特征的信息。我同时从形式和背景信息来阅读场地。比如，我曾为新泽西城市大学的艺术学院做了一个小项目。在研究中我发现这个学院的学生来自65个国家。于是我将"艺术"一词翻译为学生们不同的母语，然后在花园中创造了一个石质的环形区，人们可以坐在里面读所有不同的关于艺术的定义。

吴欣：近年来你对环境问题坦率真言。你怎样看待你的艺术和你作为艺术家的作用？与27年前设计越战纪念碑时的想法相比，想必有了很大的改变，现在的你更成熟，更自信。

林璎：我想那个时候我没这么成熟，但却更自负。长大了人会变得更有反省力和有更多的问题。年轻时，保护你的是一种对自己是对的的确信。有一种少年的天真。随着年龄的增长，我们都会丧失那份天真的自信，明白生活的复杂性。我一直是一个率直的环保主义者。当我还是青少年时，我就曾请愿停止日本的捕鲸行为，反对使用钢夹狩猎。所以我对环境的激情，很早就是在那儿了。我觉得现在中国和美国掌握着环境保护的关键。中国需要平衡环境和发展，坚信在前所未闻的经济增长和保护环境之间协调的可能性。中国人已经一次又一次地证明，他们能够在一夜之间改变很多事情。而美国则需要更多严格的立法，同时美国人必须停止浪费型的消费行为。美国拥有世界上最奢侈，也是最差的生活方式，人均碳足迹高达20吨。我现在在做的一件事情就是提倡美国人进行"碳足迹减肥"，争取每人减掉10吨。我们必须这样做。每一个人都必须尽力而为。如果我们再不关注气候和多样性问题,我们将会失去我们赖以生存的地球家园。

吴欣:所以关键的因素是减少消费?

林璎:问题比这复杂多了。包括改进和加强环境立法,清理和保护动物栖息地,减少我们的消费、土地和交通消耗。每一个人都应该帮助减少对自然资源的消耗。从少开车到少用纸和木材,从回收利用到购买和支持有机食品。如果我们每个人都改变一点,就能对保护栖息地和物种并减少碳足迹有巨大的影响。

吴欣:你的新项目"什么在消逝?"进展如何?

林璎:我刚刚在哥本哈根气候大会上发表了一个视频,是有关禁止采伐森林,以减少热量的散失,从而保护物种的多样性。它叫做"再生一棵树",你也可以从我的网站上看到。作为一名艺术家,我创作了新作品,并进行传播,从而向人们传达一种信息。它是一个多场所、多媒体的纪念碑的一部分,这个纪念碑项目同时还包括在北京艺术中心开展的巡回展览"空房"以及旧金山加州科学博物馆的永久性雕塑"听锥"。项目以概念为基础,包括视频、永久和临时装置、一个网站和一本书。在谷歌的帮助下,网站已于2010年的世界地球日正式开放。我们还利用纽约时代广场的大屏幕,从2010年4月开始,每小时一次播放这个特制的5分钟视频。这是一个有关纪念碑的实验性创意。它将会像水那样,随时随地任意流动。所以,这是一个彻底的"反"纪念碑。

吴欣:既然提到纪念碑,你能否解释一下,在你心中,纪念碑是怎样一个概念?

林璎:很多人把纪念碑当成是一个实体、静止的固状形式。通过"什么在消逝?"我在重新思考一个多场所的纪念碑。

吴欣:那你怎样看待"记忆"?

林璎:记忆和历史使我着迷。尤其是人类创造集体性记忆的能力,使历史世代相传。这也许就是为什么我一共设计了 5 个纪念碑。"什么在消逝？"是第五个也是最后一个。我打算余生致力于此。人们往往假定一座纪念碑是用来纪念过去的。对我来说，纪念碑是我们记住过去以吸取教训，塑造一个不同的未来。如果我们不能够正确地记忆我们的过去，那我们必将重复相同的错误。人类是唯一能够共享几千年历史的物种。通过历史，我们与祖先以及未来的子孙后代进行对话；通过历史，我们也许能够转变人类整体的行为和生活方式。我相信一个纪念碑是一个教育的工具。

吴欣：你把纪念碑看作人们根据现有的问题，去激活历史，以从中吸取教训，走向不同的未来的工具？

林璎：是的，它积极地与过去和未来对话；不只是消极地望向过去。这个跟"我赢了！这就是我的标记"的态度是决然不同的。历来纪念建筑都是纪念胜利的。相反，我认为纪念碑是为了失去的记忆。生活难以预测，我们应该从过去的错误和经历中吸取教训。我想这是一种非常东方化的思维。当然，我所选择展示的内容会指向某些特定的历史事实，但是它从来不是说教。有时候人们选择去遗忘痛苦的过去。然而一旦你能面对那个过去，你就能征服它。

吴欣：那你是怎样看待历史？是不是那些发生却被忘却了的事情？

林璎：人类忘却，或者方便地记住那些我们想要记住的事情。人的记忆是具有选择性的。我想有时候我们需要记住那些我们做过的不好的事情。我们也需要赞美人们是怎样对历史的发展产生影响的。就比如位于阿拉巴马州蒙哥马利市的民权运动纪念碑，赞美了人类通过牺牲自己的生命来改变历史车轮前进的方向。当你离开那个纪念碑的时候，你意识到原来单独的个人也能够有所作为，有所影响。的确，每一个人都能够产生影响，这是非常重要的。

吴欣：你刚刚提到了你的亚洲思维方式，你怎样看待自己的美籍华人的身份？你设计的位于纽约市的美国华人博物馆刚刚落成开馆。

林璎：是的，我还是博物馆的董事会成员之一，非常愿意帮助博物馆成长。他们现在在发展一个与这个实体博物馆相匹配的网络博物馆，这个新发展令人振奋。这样来自世界各地的人们，都能够进入博物馆，了解美籍华人的历史。美国人并不了解在美华人的历史。这是一段复杂而丰富的历史。很多华人参与了美国铁路的修建，被称作美国的"二等奴隶"。华人还是唯一被拒绝授予美国公民权的种族。在介入美国华人博物馆事务之前，我对这些事情也一无所知。如果忘却历史，你就会理所当然地接受现实。我有时候会想，我们是中国人么？我们是美国人么？美国人需要了解中国人在美国的历史有多深厚。尤其因为美国是一个移民国家，了解这段历史将使每个人更好地理解自己的文化识别性。

吴欣：与其他文化群体相比，华人团体没有那么主动。能有像你和发起了"丝绸之路"项目的马友友这样的华人精英很令人欣慰。

林璎：是的，华人团体往往被视为非常沉默。但我们有如此强大的群体，为这个国家的发展贡献如此之多，记住并宣传这些故事是很重要的。我想我们已经开始通过美国华人博物馆来激活这一团体。我对此越来越投入，这与20年前的我是大不相同。在过去的十年中，我有了自己的孩子，所以对美籍华人事务也变得越来越关心。

吴欣：你对中国的设计师和学生们的建议是什么？

林璎：大胆创意，但是现在从环保主义的角度来看，我们都有责任对环境产生积极的影响。我们建筑的方法，如何建，建在哪里……我们必须有可持续性地建设，对环境和下一代负责任。

吴欣：你认为当代中国设计师应该怎样对待我们几千年的古老历史和文化？

林璎：美国没能像中国和欧洲那样，拥有悠久而璀璨的建筑遗产。如果你拥有几千年的绚烂文化，请尊重它，接受它，保护它。你仍然能够修建新的现代建筑。伦敦和纽约都是新旧建筑结合得非常好的例子。有很多新锐的前卫建筑，同时整个城市有着不可思议的邻里感和历史感。在中国，如果你把一座建筑作为一个单体保留起来，围绕这个小的尺度"纪念碑式"的古老建筑修建新的大楼，比例就失调了。我们必须思考我们需要保护什么，保留什么。在我看来，中国辉煌的历史遗留下来的物质文化遗产，远比房地产重要。老街区不应该拆除。美国人犯了错误，比如丢失了纽约伟大的建筑之一宾夕法尼亚车站。反思它的毁灭，纽约市通过了《史迹保护法案》。我想中国的设计师和规划师需要放眼四周，认真评估历史遗产的价值。中国不能满足于碎片式的史迹保护，更无需拆除成片的街区。中国有平衡的能力，在新老建筑之间寻找平衡是完全可能的。（章健玲 译）

Maya LIN
—between art creation and form

Many know Maya Lin through the first work she built at the age of 21, the Vietnam Veterans Memorial on the National Mall in Washington DC. The memorial attracted public attention and stirred public debates in America, and made her to realize consciously the first time her Asian-American identity. After the debut of the memorial in May 1982, she visited mainland China for the first time with her parents in 1985, and returned frequently thereafter.

It takes courage and resolution for artists to move out of the shadow of their success; many never do. For Maya Lin, it took her more than a decade to move past the Vietnam Veterans Memorial and to be able to talk about it in the documentary film *Maya Lin: A Strong Clear Vision* (1994). On the other hand, her artistic struggle in the past 28 years — not wanting to become typecast as a monument designer — has borne fruits in a body of extremely varied new works ranging from studio artwork, to outdoor sculptures and installations, to gardens and parks, to architecture (www.mayalin.com). In 2006, she wrote her first book *Boundaries*, pondering: "I feel I exist on the boundaries somewhere between science and art, art and architecture, public and private, east and west. I am always trying to find a balance between these opposing forces, finding the place where opposites meet."

The Maya Lin I met today is confident and outspoken, a mature woman artist actively engaged with two issues concerning her most: global environment and Chinese heritage. Yet, she sees herself as "apolitical" and "nondidactic," she believes in the power of history in shaping the future. Just days before this interview, she presented a video creation "Unchoping A Tree" at the COP15 (Copenhagen United Nations Climate Change Conference 2009). This video is part of her on-going project "What is Missing?". This, her last memorial project, focuses attention on living species and places that have gone extinct or will most likely disappear within our lifetime. The message she sends to the world is that each of us can make a difference. In September 2009, Maya Lin helped to launch the new Museum of Chinese in America (MoCA) in New York City's Chinatown, designing the architecture and sitting on the board. In recent decades, she has joined other Chinese-Americans, such as Yoyo Ma (cellist) and Gary Locke (former American Secretary of Commerce and Governor of Washington State), to motivate overseas Chinese communities. She visited mainland China regularly, and her message is that Chinese designers should cherish China's rich culture and history, and balance between the old and the new.

Xin WU(WU hereafter): You have always thought of yourself as an artist and architect, how does landscape fall into that equation?

Maya Lin(LIN hereafter): Landscape is an integral part of my art and architecture; it is often the subject of my work. In my architecture I cannot separate the landscape from the architectural forms. They do not just sit upon the landscape, but they actively engage with their environs—oftentimes creating outdoor spaces that seamlessly merge with the interior spaces— they also create a frame to experience and view the landscape. In my art—the environment, topography and terrain form the inspiration and context for all of my sculptural works.

WU: So landscape is the connection between your art and architecture?

LIN: Yes—and not just formally— I think the architecture is also very committed to sustainable design solutions that respect and work with the environment. I think some of my landscape works are much more designed and one could say they are really set in a landscape design realm. However, the garden projects "Reading A Garden" and "Input" also incorporate language; they are conceptually based. These works like the Memorials use text and information embedded into the landscape. In "Ecliptic", a large scale urban park, the language I used was time—embedded in the central skating rink is a fiber optic light array based on the constellations of the night sky over the city on New Year's Eve 2000. That artwork marks one specific point in time. The park was designed in the millennium year, and the whole star pattern in the floor of skating rink marks one important moment. It adds a layer of scientific factual data as a type of textual information to order the space.

WU: Regarding the use of language, are you trying to encode space?

LIN: Yes, language is a physical material. It is not necessarily overlaid on a physical form; actually at times it is the form. One can argue that the Vietnam Veterans Memorial is pure surface; it is not so much a heavy inserted physical wall, but instead the polished surface of the earth where the names become the object. The scale of the text becomes critical—though we're used to reading monumental sized text— such as a billboard—seldom do we introduce the more personal, book-size scale in a public place—to do so inherently asks for a more personal reading of the space- one that becomes psychologically more intimate and emotional.

We have the five basic senses—sight, hearing, touch, smell, and taste—I think there is one more— reading. Reading was invented by our species; there are no other species that can read. I consider reading a sixth sense. We don't just see words; we feel words. I am very drawn to using that sensibility in my work.

WU: When you say reading, do you mean the capacity to make sense of scattered information and construct a meaning?

LIN: The act of reading is a very complicated physical and emotional act. By presenting it in these works in a very integrated way, I am seeking a way to communicate. It is not separate—I make my form and then I convey the information—I integrate the two fully. When you make a sculpture, it is pretty mute; one will react to it as it is. And then, if you say "read me" even though one is supposed to touch it, what would happen? I don't know the response from my audience; all I know is that it adds a layer of meaning, intimacy, and humanity to the work.

WU: A kind of open-ended "Input"?

LIN: When collaborating with my brother Tan Lin on "Input", both words and form evolved simultaneously. He composed a poem about memory and time and how memory works not linearly but by random connection. And I made the garden. Some people go to see "Input", and they ask how we are supposed to read it? Is there a strict path through? I would say: no, it is like memory; you may find it randomly and then you are connected. It is about our personal memory, of Ohio University in Athens, where we grew up. However, memories of a place can be universally shared; and anyone who spends time there is going to connect to different things. So literally, he wrote a poem and I came up with the form. Then I started manipulating the location of each component to fit his words. How the words and their meaning influenced the spatial arrangement that I came up with was quite illuminating.

WU: How about "Reading A Garden"?

LIN: It was the first time we collaborated and it was pretty much the same. I made an initial concept of a garden. He wrote the poem. I split it into three parts that turned into three pathways going into the garden. He went back to manipulate the poem…to relate to the physical pathways.

WU: In comparison to "Input", the poem was more visual.

LIN: Yes. It was more playful. We wanted it to have a quality almost like "Alice in Wonderland." Not necessarily for children, but we wanted the words to be accessible and inviting. For me, the whole garden is about the directionality and weight of words.

WU: That was a brilliant project. The letters invite reading but in a totally unfamiliar way of engaging with words.

LIN: Yes, absolutely. The concreteness of the words and poetry has a physical draw that is being played out. In poetry, sometimes trilogy series exist; my brother and I are waiting for a third opportunity in Ohio, after "Reading A Garden" (Cleveland, Ohio) and "Input" (Athens, Ohio), to complete our trilogy. Then it will be published in book form, to honor my parents and our family life in Ohio. So it would also be a trio made of garden, poem and book.

WU: That's wonderful; you wanted to celebrate your up-bringing.

LIN: Yes, there are fond memories. Both my parents are scholars from China — my father a ceramicist and my mother a poet. They taught my brother and I a lot, and allowed us to develop our voices. I went into art and my brother went into writing. Of course, we followed in our parents' footsteps…

WU: In history, Chinese artists have always strove for the "Three Perfections":

painting, poetry, and calligraphy, all integrated in one frame.

LIN: Interesting. As children, like many Chinese Americans who grew up during the Cold War, we were not taught the Chinese language, and we were never exposed formerly to Chinese art. The idea of the "Three Perfections" is new to me but if I think about it, the image, language and calligraphy are beautiful and I am drawn to language and to form making simultaneously.

WU: What you do is certainly very different, but there is an aesthetic attitude that seems very Asian to me, since you attend to things in a context of relationships rather than in isolation.

LIN: Surely, I can't isolate things; I have to see them in a whole. Take art, architecture and memorials, I see them as a tripod. I cannot take away any of the legs. I have worked more in art and memorials than in architecture. To really develop my architecture it will require more time to produce enough work— Maybe another decade since I can only do one project at a time. I am very committed to Architecture as an art form.

WU: Would you consider yourself a landscape architect?

LIN: I really love working in the land—and creating environments. But the gardens are not like the project "Wavefield" at Storm King Art Center, where the form is pure art. If one puts a pathway through an artwork to turn it into a functioning space, it can destroy the purity of the art form. I am definitely attracted to making landscapes, but I am in a slight conflict. Part of me is really drawn to the sculptures and the earthworks that would be destroyed if they were made functional. So, even though I love making urban landscapes, it might be better not to add one more discipline to what I do. My work has always been difficult to be categorized.

WU: That happens to all people who are crossing boundaries. How do you deal with scale when working in different domains?

LIN: My works are fairly large whether they are architecture, parks, or sculptures yet they are always very human, never monumental. Photographs never convey the idea of the work. You have to experience these works to understand them. I could only think of my work as how one person walks through it, touches it and feels it. It is exceedingly psychological. No matter what the pieces are, they are about the private viewer's response to the art.

WU: So it is the experiential that is important to your work, not the visual?

LIN: Absolutely. I always say that they really don't become works until someone walks through them and experiences them. Robert Irwin was very influential to me. I thought Irwin was a great artist after reading *Seeing Is Forgetting the Name of the Thing One Sees: A Life of Contemporary Artist Robert Irwin* (by Lawrence Weschler, 1982), though I had never seen his works. Early in his career, Irwin made some incredible disk paintings but forbade them to be photographed. He forbade photography because the work explored the edge of perception and photographic images could not tell anything about them.

I am interested not only in hybrid works, but by the threshold of perception, which implies a subtlety that demands attention in order to perceive the artwork. When I finally saw one of those disk paintings years later, I realized why Irwin did not want this work to be photographed. You could never capture the true experience of it. This is also true for James Turrell, whose work has to be experienced first hand. Those two artists have been very influential to me. I find the experience and idea of art almost on the edge of perception to be fascinating.

WU: Can you talk about your approach to design on an actual site?

LIN: I visit and research the site very well. I go many times, see it, absorb it, dwell on it, have it in my head, and record the existing conditions in models. Then I put it aside and research everything relevant that I want to learn. For example, with the first "Wave Field" (Ann Arbor, Michigan), I studied aerodynamics for a year before I came up with the design. I study scientific information, history, time, and language. I meet with people, but depend more on research. And then I disappear, I go into my studio and I make something. Generally, what I have created in the first place is very close to what is executed. In order to discern what I have to do on a particular site, I begin to understand it on a conceptual level through research. Design is a process of thinking and researching to me. This process can take up to 6 months or a year. During the research I try not to design anything, not even to put the pencil to paper. When the research is done, I put it away and allow the idea to come. It is crucial that I try not to go to the site with a preconceived idea.

WU: In your book, *Boundaries*, you said you try not to reach the form too soon.

LIN: Right, otherwise you stuff whatever you want to do into the preconceived idea and form. To me, design is a much more interactive process when the creation balances the research.

WU: How do you know when the right form arrives?

LIN: I just know. You know when you are done. Though the hardest part is to figure out what I am going to do. I am extremely site-specific. With every new work, I literally have to reinvent my voice to fit the site. And that is hard. I am always worried that I may fail to come up with what I want to do. I generally don't accept commissions with a deadline, since I might produce something because of the deadline and not because of a natural discovery process. The form in my work is first of all an idea with a specific materiality.

WU: While researching, do you focus on certain aspects?

LIN: I start exploring without a specific direction, seeking information about not just the physical but also the cultural context of a place. I read the space for both formal and contextual information. For instance, I made a small sculpture installation for New Jersey City College School of Fine Arts. In my research I discovered there are students from over 65 countries attending the college. Then I took the word "Art" and translated it into all the languages reflected in the student body. I then created a stone circle in a garden that you can sit in and read all those definitions.

WU: You have been very outspoken about the environment in recent years. How do you envisage the role of your art and of yourself as an artist? I suppose it has changed quite a bit since the Vietnam Veterans Memorial 27 years ago, now that you are more mature and certain of yourself.

LIN: I think I was less mature but more sure of myself then. As you get older, you get more reflective and possibly question more. What protects you when you are young is the belief that you are right. There is a naivety of youth. One loses that naïve certainty when getting older. You understand life is more complicated. I have always been an outspoken environmentalist. When I was a teenager, I would petition to stop whale hunting in Japan, and to ban hunting with steel traps. So in terms of my passion about environment, it has been there from the start. I think right now China and the US hold the key. China needs to mediate environmental protection with a belief that it can balance between a type of growth that the world has never seen and a need to protect the environment. The Chinese have proven again and again that overnight they can change. Americans have to legislate themselves to much tougher laws, but Americans also have to stop wasteful consumptive practices. America has the greatest or worst lifestyle in the world, and a huge carbon footprint of 20 tons per person. One of the things I am going to do is to suggest that Americans go on a "carbon diet" to lose 10 tons each. We have to. Each one of us needs to do as much as we can. If we do not pay attention to the climate and biodiversity issues, we will destroy the very planet that sustains us.

WU: So the key factor is reducing consumption?

LIN: It is far more complicated than that. It involves improving and enforcing environmental legislations, cleaning up and protecting habitats, and reducing our consumption and land and transportation use. It is important to show how each one of us can help to reduce wasteful consumption of our natural resources. From driving a car less to using paper and wood products, from reclaimed or recycled sources to buying and supporting organic and sustainably managed food products. If each one of us can change we could make a huge difference in helping protect habitats and species and reduce our carbon footprint.

WU: How is your new on-going project "What is Missing" coming along?

LIN: I have just presented and launched a video, at the Copenhagen Climate Conference, about preventing deforestation as a means to reduce emission and protect species. It is called "Unchopping a Tree", which you can see from my website. This is part of a multi-sited and multi-media memorial, which also includes the "The Empty Room" traveling exhibit that opened at Beijing Center for the Arts, and the permanent sculpture "Listening Cone" at the California Academy of Sciences in San Francisco. The project is conceptually based, with videos, permanent and temporary installations, a website, and a book. With Google's help, the website is scheduled to be launched on Earth Day 2010. We will also take over a billboard in Times Square with a five-minute video piece running for the month of April every hour. It is a very experimental idea about a monument. It will be just like water, flowing wherever and whenever it wants. It will be a quintessential anti-monument.

WU: Since you state your position with respect to monuments, can you clarify your concept of a memorial?

LIN: Well, most people see a monument as a physical and static object with a solid form. For "What is Missing?" I am re-thinking a memorial to be a multi-sited work.

WU: What do you think of memory?

LIN: Memory and history are fascinating to me. And what I am drawn to is our unique ability to create a collective memory with history, so that we can remember from generation to generation. This is perhaps why I have created 5 memorials in my lifetime of which "What is Missing?" will be the 5th and last of the memorials—I thought I will work on it for the rest of my life.

I think there is an assumption that memorials are there to commemorate the past. For me a memorial is a way for us to remember the past, so that we can learn from our past and shape a different future. If we cannot remember our past accurately, then we are doomed to repeat the same mistakes. Human beings are the only species that are capable of sharing thousands of years of history. Through history, we communicate with both past and future generations. Through history, we can help shift how we act and how we live. I believe a memorial is a teaching tool.

WU: So you see memorial as a device that reactivates the past for concerns of the present and leads us to a different future?

LIN: Yes, it is proactive and engages the present and future— it doesn't just passively look at the past. This is very different from the attitude that "I won, here is my marker"—if you think of how the commemorative architecture has been, it commemorates a victory. To the contrary, memorials are about remembering what we have lost. Hopefully because life is unpredictable, we are learning from our mistakes and from what we have just gone through. I think this is a very Asian way of thinking. Obviously, what I choose to present directs you to certain facts in history, but my works are never didactic. Sometimes people choose to forget the painful things in the past. But if you can face that past, then can you overcome that past.

WU: So what is your view of history? Things happened that we forget?

LIN: We forget or we conveniently remember what we choose to remember. We have selective memory. I think sometimes we need to remember and point out the bad things that we did. And also we need to celebrate how people can make a difference. Like the Civil Rights Memorial (Montgomery, Alabama), we celebrate how individuals change history and the sacrifices they made. So you will walk away with the idea that one person can make a difference. Again, one person can make a difference; this is hugely important.

WU: You just mentioned your Asian way of thinking, what about your Chinese American identity? You designed The Museum of Chinese in America (New York City) that opened recently.

LIN: Yes, I am on the board and am very interested in helping them grow. Their new direction is exciting. They are making an on-line museum that matches the physical museum, so that anyone, from anywhere in the world, can check in and understand Chinese-American's history. Americans don't know the history of Chinese-Americans. Its a complex and rich history—We were called the "Second Slaves" of the country because of the railway. We were the only race to have been singled out for exclusion from citizenship. I had no idea of all these things before my involvement with The Museum of Chinese in America. So again, if you forget history, you take the present for granted, especially with China and US. Are we Chinese? Are we Americans? Americans need to understand how deep of a history the Chinese have in the US. Especially since America is a country of immigrants— to understand this history will give everyone a better understanding of cultural identity.

WU: The Chinese community has been far less forthcoming in comparison to other cultural groups. It is encouraging to have people like you and Yo-yo MA who launched the Silk Road project.

LIN: Yes, the Chinese community is perceived as very quiet. But we are such a strong presence and have contributed so much to growing this country and it is important to remember and tell those stories. I think we have started to motivate the community with the Museum of Chinese in America. I have become increasingly involved, which was not the case 20 years ago. I think in the last decade, while having kids, I have grown much more interested in being involved with Chinese-American issues of identity.

WU What is your advice to Chinese designers and students?

LIN: Take risks in your creation, but right now from an environmentalist point of view, we all have a responsibility to have a positive impact on the environment. The way we build, how we build, where we build. We must try to build sustainably and with a sense of responsibility to the environment and to future generations.

WU: How do you think contemporary Chinese designers should deal with the history and culture of thousands years?

LIN: America doesn't have the privilege of having a splendid architectural legacy like China and Europe do. If you have a brilliant culture that goes back thousands of years, respect it, embrace it, preserve it; you can still build new modern works. Look at London or New York, both excellent examples of the mixing of old and new. There are great new innovative architectural works, but there is also an incredible sense of neighborhood and history throughout the city. In China, if you protect a building as a single entity from its neighborhood and you get new buildings surrounding the smaller historic architecture, you loose the scale of the places. We have to think about what we are trying to protect and preserve. To me the heritage from China's amazing past is far more important than real estate. Old neighborhoods should not be torn down. America made mistakes, for example, losing the Pennsylvania Station one of the great edifices of New York City. Reflecting upon its destruction, New York City created the Historic Preservation Act. I think that Chinese designers and planner should look around and seriously assess the value of their historic heritage. You cannot save piecemeal; you don't have to tear down everything. You have the ability to balance; and it is possible to have a balance between what is old and what is new.(Translated by Jianling ZHANG)

帕奥罗·伯吉
——景观设计与创造性的诠释

瑞士景观设计师帕奥罗·伯吉在欧美景观设计界颇为知名,不仅是作为一位思辨的设计师,同时也是一位极具启发性的设计教授。他的项目"卡尔达达:对一座山的再思考"(Cardada: Reconsidering a Mountain)是 2003 年"罗莎·芭芭欧洲景观奖"得主(我非常高兴伊利诺伊大学的戴维·海斯(David Hays)教授特别为本次的访谈就这一项目写了一篇评论)。他同时还任教于意大利威尼斯建筑学院和美国宾夕法尼亚大学。伯吉设计中的形式往往表现为"极少化":一条山间小径,一个林中挑台,一垄彩色的田,或者仅仅是一圈树……然而他原创性的根源却在于对世界万物极其丰富细致的感受和浓缩——"做一个好奇的观察者",伯吉在访谈中屡屡强调。这种态度与中国人所熟悉的新儒家的"格物致知"认识论相通。所不同的是,伯吉通过观察所寻求的不是"理",而是感性体验,是一种超越视觉的、超越想当然的、表象的、人与自然的交流。在他最近给我的信中有关苏州园林的一段话能帮助读者更好地理解其想法:"我不禁想起去年有幸参观过的园林。我感觉好像身在其中,大概因为此时外面正在下雨,跟我游园时一样。我尤其记得环秀山庄。那是不可思议的感觉,走在山石间有一种不可思议的感性体验:你会经历攀登一座真实山峰时所有的妙处——那种发现的快感、那种惊奇、那种愉悦和敬畏、那种害怕掉下去的焦虑……多么不可思议的设计!我一定要回去再体验体验。"欣赏伯吉的设计意味着一种对周边事物的再了解,以全部的感官去体验,以好奇心重温山水。但最重要的还是,懂得设计是一种当代性的、以特定地点的自然和文化的历史为启迪的、诠释性的创造过程。

吴欣：卡尔达达的高山大湖是中国人概念中完美的山水，我本人和很多其他人都是最先从这个项目了解你的，因为访谈后有一篇关于此项目的专门评论，这里就不再赘述。不过这是一个非常的作品，加上你在设计该项目时曾有过一位中国助手，我想还是以它来作引子。在参观卡尔达达时，我印象最为深刻的是你的设计力图倡导并且引发人们去享受和体验山水。在项目之初，这就是一个核心的设计理念么？

帕奥罗·伯吉（以下简称伯吉）：是的，我想是的。卡尔达达项目是由一个非常简单的问题开始的。我们不得不重建城市通往山上的缆道，这是一个非常昂贵的过程，所以引发的问题既简单又极其关键：人们由新的缆道到达那里，发现那儿一如往昔，更像是一片城市的边缘地带，而不像是能融入自然的地方。这是根本问题。这个项目就像一个游戏一样开始了——没有任何的计划。没有人能告诉我应该做什么或者知道我每天要完成什么。我做设计的同时也做规划。我要思考哪些地方更重要，哪些地方是要改动的，并寻找恰当的切入点。一方面，它是一个简化和净化的过程，排除所有不当的干扰因素；另一方面，它不但应该彰显场地的自然美，也必须揭示视线所及的风景本身的历史。以前当地人去卡尔达达就像去其他所有地方一样：他们只是去那里放松，看看山，然后就回家了。就这么简单。我想要提醒人们在我们极目所见的地平线背后还有东西，想让人们变得更好奇并敢于询问"在地平线的背后还藏着什么东西？"这种信念在整个项目的进展过程中自始至终驱使着我。

吴欣：能详细解释一下你提到的"风景本身的历史"这个概念吗？

伯吉：我注意到那些惯于去这座山的人，总会去一个地方坐着看远景。我是一个非常好奇的人。如果我对什么事有疑问，就想了解更多。我想用设计传达出这种好奇心，那种作为一个人想要了解更多的感觉。比如，当你来到卡尔达达，映入眼帘的是马焦雷湖（Lago Maggiore）和湖中的小岛。这些岛屿很美，很可爱。但实际上，这些岛屿和住在那儿的人都有着非常有趣的历史。我们发现岛的第一个主人是一位俄国公主，她邀请了许多宾客。其中一位是伟大的作曲家弗朗兹·李斯特（Franz Liszt）；他是她的音乐老师。所以在设计的地质观景坪周边，我们列出与这些岛屿有关的图片和文字以激发好奇心，却不提供任何答案。比如，一个游客看到弗朗兹·李斯特的名字，可能会想"李斯特和这些岛屿有什么关系呢"。尽管这只是一个小小的细节，但我们就是用这种方式来激起人们对隐藏在视野背后的故事、对每一个特定的景点和许多你眼前景观背后不为人知的历史的好奇心。每一个事物背后都是如此地丰富，对于我来说，这就是景观设计应该试图传递的信息，使人们的思绪在时空中游走。

吴欣：这个项目有两个特别的地方。首先，你的办公室就在附近，能真切全面地了解那座山，并且从一开始就参与在项目的过程中。其次，我知道你是一位飞行员。在我看来，这两个方面的优势使你对卡尔达达的景观有一种独特的了解：你既可以从高空俯瞰它，又可以就近体会它所有的细微之处。这两种截然不同的视角是怎样成就你的设计的？

伯吉：从来没有人问过我这个问题，很有趣，好像我有坐拥感知优越的"嫌疑"。或许这种优势是有的，卡尔达达项目前后经历了5年，我亲自测绘了山上的许多地方。但设计主要还在于感知的方式和途径。近水楼台自然有利，但离场地很远也可能有帮助，你会有非本地人的视角。这样的视角可能要开阔得多；一种好奇的观察者的目光，不会习以为常、不以为然。在最近一个研究德国城市农业用地美学的项目中，我们在场地上乱逛着看野花。这和场地及它的历史有着非常密切的关系。使我们的设计具备一种尊重在那块土地上耕耘的农民的视角和观点。

吴欣：我们稍后再回到这个城市农业的项目。刚才你提到一个非常有趣的概念——设计者是对场地的好奇观察者。当你碰到一个新场地，你的切入点是什么呢？

伯吉：观察一个场地可以有许多切入点。或许第一种方法是这个地方的地貌、结构、地势等。但然后要考虑很多其他因素。比如说，光。光能发挥很重要的作用。当然，然后是历史。如果你在这个地方，你了解这个地方的历史，那么这就是一个能够传递其他信息的切入点。如果是一片空地——比如说一马平川的农田——或许你会发现几百年前在这个地方曾有过一场战役,那么这片田就变成了一个不同的场所。所以这是一个关系到感觉、体验和情绪的问题。你瞬时领会到的一切都可能带来对设计的某种融入。

吴欣:设计评论经常把你的设计和极简主义联系在一起。你对此有何见解?你创作的是极简主义作品么?

伯吉:是有极简的成分,但是这必须从少就是多的层面上来讲。设计是一个过程,我的创作从来不是极少,而是极多:我先尽情地感受,然后通过研究力求去除所有多余的成分来达到强烈的表现力。更确切地说是一个提炼的过程;精炼只能来自于丰富。这个道理不光适用在建筑上或景观上,在音乐和文学中也一样,比如诗歌。能摒除一切无关紧要的东西是相当大的成功。这非常难,需要长期的训练。我喜欢简约的艺术、简约的写作、简约的设计;因为它们能表达强烈的感受,是浓缩的、有冲击力的。

吴欣:现代生活往往使人因太多东西而感觉负担过重。你想用设计来尽力清除一切不重要的东西,来帮助人们重新和自然对话?

伯吉:我的确认为可以在每个项目中传达你看待环境的方式,那是种更加关注、敏感,甚至是种对环境恭敬的方式。作为一个设计师,你需要具备简化的能力,但尽力去传达一些信息也是非常重要的。为了传达信息,你需要一种灵感。组构一幅画或是一个项目都来自灵感,来自想传递信息的渴望。当然,美观也非常重要。比如在卡尔达达的项目中,我们注意到至今尚未有任何的人为损坏。如果某人真的喜欢某种东西,他是不会破坏它的。如果你注意到了那些人们喷画和涂写的墙面,那上面常常不会传达什么美的信息。

吴欣:我想知道你怎样看"美"。这好像是现在很多设计师害怕用的一个词。比如你在卡尔达达附近的一个新项目"佩科西花园"(Giardino dei Percorsi)。

伯吉：佩科西花园是一个非常小的私家花园。先提一句细节，我所有的花园都有自己的名字，像中国园林的命名那样，不以业主的名字命名。这个花园的名字的意思是"万径游园"，因为设计的主题就是一个交错的步道系统，邀请你以无穷的方式在花园里漫步。这座别墅位于背山面水的高坡上，开始的时候花园里只有一个离建筑近的地方可以供人坐。我的设计探索了可以穿梭于这个空间的多种可能性。你穿过草坪的方式很重要，因为它建构起了视野、湖泊和岛屿的关系。我觉得如果你可以步行穿梭其中，在里面体验不同的视角，并且还可以停下来坐坐某些地方的话，那将会是一种独特的享受。抑或你来到花园的边上，伫立在那儿，你可以看到某些事物———一扇特有的景观之窗———或者你看到树林里的一棵奇特的树，或是其他东西。我们还将花园扩展到悬崖上，种植了一些可生长在岩石上的奇花异草。随意地漫步于自然之中的体验正是花园的所在。这就是我所理解的美。它不是关于某种形态或模式，而是关于感知和发现的体验，如何成为一个好奇的观察者的体验。那是很美妙无穷的。那就是我所想象的花园。

吴欣：在这个设计中，你想传达的是什么？

伯吉：多种不同的氛围。比如，远眺地平线时的那种惊喜；或者在花园中穿行的愉悦；或者进入空间的不同方法。无论你在哪儿，都有驻足片刻、观看欣赏的诱惑。

我告诉过你关于我最近在中国的经历。我在20多年前就到过苏州，记忆犹新。去年访问时说想再去看看这些园林。有人警告我："如果先前的印象太好了，再看一遍是会了然无味的。"而我不但称赞新的园林，还包括那些以前参观过的。故地重游，我不但没有改变观点，相反，在仔细琢磨后反而更加喜欢它们，理解它们的丰富性。网师园是我特别喜欢的花园之一。它是这样一个花园，我该怎么说呢，永无穷尽。你永远都不可能全部体验完它，总是有一片你从来都没有到过的风景在眼前展开。你可以一直走，一直走，一直走……它是一个永无止境的发现过程。

我欣赏中国园林，因为这些艺术品只能是由长期深入洞察的观察者创造出来。另外一个园林（环秀山庄）非常小巧。你第一眼会看到一个由岩石组成的中央空间，它被一些水池分隔。当你开始进入其中时，你会感觉像爬山一样——不是一座小山，而是一座高山。一个非常出色的设计，它传达给你攀登时所有的真实体验。为了能够创造这样的作品，你必须得是一位杰出的观察者；不仅如此，你必须能够塑造出这种感觉。许多人都有过登山的经历。但是如何体会这种感觉并把它用形式表达出来，这是设计，是艺术。你知道不可思议的是什么吗？这些园林虽然已经有几个世纪之久了，它们仍然有着不朽的能量。很多设计——那些脆弱的形式——只能展示片刻的靓丽光彩，其表达的感性体验却是短暂的、经不起时间考验的。而艺术是永久的。

吴欣：把园林解读成"赋形于体验"很精当。这正是很多人在游览苏州园林时忽略了的，他们只是看到了表面的形式而没有看到形式背后的体验。

伯吉：去年到中国时应邀在南京作报告。我解释了我的哲学——好奇心和我关于中国艺术的想法——很多听众上来告诉我说他们和我有那么多共通的东西。能产生共鸣使我感到很骄傲。当和中国教授们一起去参观苏州园林时，我非常陶醉。我们像欧洲人欣赏艺术品一样观赏这些园林——不仅是在游览园林，而且是在解读园林。有那么多的事物来帮助你思考：走、看、闻、读、听……这是一种全方位的体验。

吴欣：广博的人才能成为一个好的设计师。

伯吉：完全正确。我对很多事情——每天发生的事情都很好奇。我可能从报纸上的一句话得到灵感，为能回答你的问题而高兴……一位加拿大钢琴家格伦·古尔德（Glenn Gould）的故事使我感受很深：他旅游到了——大概是在美国——高速公路的某个地方，在一个路旁餐馆进餐。这种情况在全世界哪里都会发生，没什么特别的。在他周围是几桌素不相识的旅客，边说边吃，都是很平常的事。慢慢地古尔德开始关注这些人都在谈论些什么，一桌接着一桌，迥然不同。回到工作室，他用这个经历创作了一首有着不同旋律的曲子。高速公路旁的餐馆是多么的平淡无奇,而他却把它作为想象和创作的源泉,把那种体验转化成美妙的音乐。我们为什么不能把类似的日常经验转化到建筑和景观设计中呢?只要是一个好奇的观察者,体验的灵感是无处不在的。

吴欣:你的"伽斯瓦涅公园"(Parco di Casvegno)项目跟音乐就很像。

伯吉:是的,这个公园项目开始于25年前。项目背后有一个非常有趣的故事。场地本身是个公园,稀疏地种着些树,没什么设计,不过面积很大。拥有这个公园的精神病院想把它对公众开放,不过一般情况下人们是不愿意来一个精神病医院里的公园溜达的。于是我设想了一个远离医院建筑的公园。在公园里我想用树木设计一些"Follies"。这个项目的成本非常低廉,除了树什么也没有。这个"Folly"的想法其实有点像开玩笑。"Folly"的意思是"疯狂",也是欧洲传统花园中一种装饰性的亭子式构筑物。我心想"应该会允许在精神病院用树木来干这样疯癫的事情吧?"然后我和一位精神病医生一起吃了个午饭,发现他比我还要疯狂!所以,我决定就照打算的那样去做。我们曾计划每年都在这个公园里加一个树亭。我必须说我很惊讶自然的伟力。

吴欣:这个项目现在是不是一直还在进行着?

伯吉:是的,在这个公园里总共有七八处不同的树亭。这个数字还在不断增长中。可惜其中一个设计已经被破坏了。

吴欣:当初去你的工作室时,我很惊讶你原来还拥有一家苗木基地。植物学知识在这些树亭的创作中显然至关重要。

伯吉:我也做过只有混凝土而没有植被的项目。混凝土有它自己的美。不过,多年以来我渐渐地从项目中植物的身上发现了越来越多的意义。一棵树不仅仅是你种植的东西,树有历史,含含义。一棵树的意义因国家不同而异。有些树喜欢孤立生长,有些可以紧紧地一个挨一个地种植生长在一起,另外一些树之间也会有竞争。在1985年,我种了一圈杨树,这是一项包含了100棵杨树的大地艺术的一部分。当初种植这些树时，它们大小都是完全一样。现在它们成长了25年，各自迥异。有的健壮，有的瘦小，一些染了病，一些已经死了。就像人类社会一样。这就是生命本身。

吴欣：你之前提到了在德国埃森（Essen）的城市农业项目。能介绍一下吗？

伯吉：整个项目开始于3年前，景观设计用了2年时间，现在已经完成了。这是一个不同寻常的设计介入，旨在展现农业景观，问题是"农民能在种地的同时兼顾美吗？"这是个非常具有挑战性的新型的问题。在历史上很少有例子能够作为参考，而像这样的项目实际上从来没有人做过。对于我们来说，它是个有着许多不同设计的研究项目，它要分享那种传递美的共同的渴望，而不是农作物值多少钱。从在田垄间种植不同颜色的野花开始，我们都尽量使美丽和实用结合。

这个项目的设计极其微妙，在有无之间。你在这种景观里穿梭，你看到一种设计，但你会问你自己："这是人工的吗？它必须要这样吗？为什么他们这样做呢？这是随便做的吗？为什么会有一排花在那儿呢？"它没有告诉你答案，而是将这些疑惑留给了你——就像卡尔达达项目一样。设计激发好奇感，使市民带着一种新鲜感觉来看待这片农田。另一个重要点是，我们应该记住农业景观是辛勤劳作的结果。为了把这一点突显出来，我们研究了农民是怎样用机器耕作的、怎么犁地并怎样用一种美学的方式把这些表现出来。

吴欣：那片地有多大面积？都种了哪些作物呢？主要是从美学角度来挑选它们吗？

伯吉：面积大概有 $100hm^2$，非常大，是中央公园的规模。我们选用了不同种类的花卉，间种在现有的农作物之间，设计看来很小、很精致，所有的细节都需要大量的调查研究。我们请教农民一系列的问题，学习和理解他们是怎样工作的，包括何时是收获的最佳时节、什么时节筛选、怎样耕种等。弄懂了这些问题之后才能开始去做设计。成排的花卉将在不同的时节在麦田间绚烂绽放。

吴欣：还有人耕种这片土地吗？

伯吉：当然，农业生产正常进行。这不是过去老一套那种农田变公园的假设计，而是将公园引入到真正的农田里。是要使市民们发现并接受农业景观中辛苦劳作所产生的美，是一种全新的公园概念。农民从公园管理处得到一些补贴，协助花卉维护，同时，他们的耕耘又降低了市政维护的成本。对所有相关人群来说，这都是一个挑战性的项目，要求我们以新鲜的敏感性去感知环境，做一个好奇的观察者。设计在此传递的是对自然的感性体验，如季节和生长。

吴欣：这个项目是城市公园和农业生产绝佳的结合，较之现在许多城市的做法高明许多，时间允许的话真希望了解全部的细节。显然作为好奇心和观察欲使你的设计受益匪浅。不过这样一种从点点滴滴中获取灵感的能力自然不是一朝之功。你怎样训练你的学生？

伯吉：我教授给他们我的感觉和观察，没有任何诀窍。老师可以强调自己的原则，也可以鼓励学生们不拘一格，勇于创新。我赞成后者。在许多设计学院，大多数学生可能都很擅长研究、构思、绘图、观察和阐述。但塑造感受的那一步对于所有学生来说总是最难的，也就是我说的"赋形于体验"。对于我来说，很好地完成这一步恰恰是个巨大的挑战：也就是，我们怎样和为什么要由抽象的感知转向具象的形式。当然，这也是整个设计创作过程中最美妙的时刻。

吴欣：你会建议青年学生们用某种方式进行自我训练吗？

伯吉：当一个学生到了这个时刻时，老师可以陪着他一起创作。我觉得这种方式一旦进入学生的脑海，他们就能开始用这种方式工作了。比如，在意大利教学时，我发现学生们很想了解历史背景知识。而恰巧正是这种知识有时沉甸甸地压在你肩上以至于你可能失去了创作的能力——甚至是创造性。

我觉得学生能够放开思想，从原则中跳出来是非常重要的。你应该关注任何事情：音乐、艺术、建筑、政治，所有的事物。这种好奇心能够使你丰满，使你能够把它们融入到你的项目中去。实际上，源自于你的想法，并继而通过你的胳膊、你的手落实到纸上的图纸是一种长期研究和观察的结果，但这些过程是不能被建构的。这是时有时无的。知识全面非常重要。我发现现在只有一少部分人对知识好奇。我在欧洲和美国教学时注意到在许多年轻人中，他们的一些基本知识很薄弱。创造性的过程中需要这些背景知识，比如绘画。即使你不能把一个知识体系全面地和另一个联系起来，知道有联系本身就很重要。关键的问题在于如何通过设计来重新诠释体验感受。

吴欣：你刚才指出了一个很重要的问题：再诠释作为创造的一种模式。你在欧洲和美国任教，并且也访问过中国的大学。能分享一下你对于景观设计教育的见解吗？

伯吉：当然。我所说的再诠释不是就形式而言，而是就感觉和体验而言。你也可以说这是一种转译。我相信这一点对中国设计师来说至关重要。中国拥有如此多姿多彩的文化，包罗万象。如果我在中国做设计或者教学的话，我的重点会放在重新诠释和转译上。怎样充分而完全地阅读和感知历史，然后用现代的语言来表达出来，这是关键的挑战。它不仅仅是复制形式和描述，而是对价值的再诠释。要用最最尊敬和接受的心态来看待本土的文化和历史。景观设计是一个拥有无限创造可能的开放领域。

需要教授学生感知的能力，和如何把它融入到设计中。就我们的感知而言，没有任何一样是前无古人的。设计的难点在于怎样通过形式来表达它。我喜欢和学生一起工作。他们有能力，但仍需要指导。教授迫使自己进入学生的世界是很重要的。我在威尼斯、费城和南京都见过学生对场地做出了很好的研究。难点是怎样把发现的东西转化到项目中。设计不仅仅是分析，还是一门艺术。我鼓励我的学生大胆创新，不拘一格，发散思维，开动大脑。不要急于求成，要反复检查自己的观点，批评自我，时刻准备着颠覆一切。而最重要的是，做一个对你周围世界好奇的观察者，观察自然、观察文化、观察历史、观察社会，这些才是真正的创造之源。 （张夕 整理，李颖 译，田乐、吴欣 校）

Paolo BÜRGI
—between design and creative interpretation

Swiss landscape architect Paolo Bürgi is well-known in the field of landscape architecture in Europe and America, not only as a thoughtful designer, but also as an inspiring teacher of design. He is the winner of the 2003 "Rosa Barba European Landscape Award" for the innovative project — "Cardada: Reconsidering a Mountain", about which I am glad that Professor David Hays has accepted to write an exclusive review for this interview. He is an adjunct professor of landscape architecture at the Istituto Universitario di Architettura di Venezia (IUAV) and University of Pennsylvania. Designed forms in Bürgi works often take the minimal appearance: a mountain path, a forest terrace, a colorful furrow, or simply a ring of trees...Yet, the root of his creation lies in his maximum effort in drawing and distilling inspiration from the myriad things around — "Be a curious observer of the world" as he insisted repeatedly in this interview. Such an attitude certainly calls to mind the neo-Confucian epistemology of "ge-wu zhi-zhi" in Chinese culture. Instead of pursuing for li, the principle, Bürgi's observation strives for the experience of the senses, for a communication with nature beyond merely the visual and the taken-for-granted. The following words from one of his emails to me about the yuanlin of Suzhou may help us to understand him better: "I cannot avoid to think back to all the wonderful gardens I had a chance to visit last year. (I feel as if) among them right now, maybe because here it's raining like it was when I visited. I particularly remember "The Mountain View with Embracing Beauty". An incredible feeling since walking between its rocks was an incredible experience for the senses, where you could live all the impressions you may find in the mountains: the enjoyment for discovering, the surprise, the sense of pleasure as well as the fear, the anxiety when you are scared of falling... What an incredible masterwork! I know I have to return." To appreciate Bürgi's design is to dwell the world afresh, to experience through all senses, and to relive the mountain-and-water with curiosity. But first of all, is to understand design as contemporary interpretations nurtured by inspirations learnt from the history of both nature and culture of a specific place.

Xin WU(WU hereafter): The mountains and lakes at Cardada are the prefect embodiment of the Chinese concept of landscape, shanshui. Since there is an exclusive review about the Cardada project, I shall not dwell too much on it in our conversation. However, it was an unusual project for you, plus one of your assistants in the project was Chinese, I would like to use it as a prelude. When I visited, one thing impressed me most was that your design strives to bring up and entice people into the enjoyment of experiencing a mountain. Was that a core concept at the inception of the project?

Paolo BÜRGI (BÜRGI hereafter): Yes, I believe so. Cardada was a project that began with a very simple question. Because we had to remake the cableway from the city to the mountain, this was a very expensive process and the question that arose was simple but fundamental: people arrived in the new cableway and found the same mountain like before, more like a periphery of a city and not like a place where you have the feeling that you can merge with nature. This was a fundamental issue. The project began like a game — there was no program. Nobody said what I should do or knew what I was expected each day. I did the project, also the program. I looked at which places were to me more important; where I would like to intervene, and at which specific spots. On the one hand, it was a process of simplifying and cleaning; you take away all disturbing interventions. On the other side, it was an attempt to reveal the beauty of the site but also the history of its horizon. People used to go to Cardada like they go anywhere else: they just go and relax, look at the mountains, and then return home. It was a very reductive approach. I wanted to remind people that there is something behind what we see, to make them a little bit more curious and willing to ask "what is hidden behind the horizon". This was the feeling that had been driving me throughout the whole process of this project.

WU: You used the term "history of the horizon". Could you explain more?
BÜRGI: I noticed that for people who were used to go to this mountain,

there was a place where they sat to enjoy a kind of panorama like view. I am a very curious person. If I see something, I have questions and I want to know more about it. I wanted my design to transmit this feeling of curiosity, of being a person wanting to know more. For instance, when you come to Cardada, in front of you, there is the Lago Maggiore and its islands. The islands are nice and lovely. But in fact, there is a very interesting history behind these islands and about the people who lived there. We found out that the first owner was a Russian princess and she had many guests. One of them was Franz Liszt, the great composer; he was her music teacher. So on the designed Geological Observatory, we provide images and names related to these islands to create curiosity, but without giving any answers. A visitor sees, for instance, the name of Franz Liszt and he may ask himself what Liszt has got to do with these islands. Although this is just a small detail, we work in this manner to provoke curiosity towards what is hidden behind the horizon; towards each particular spot and the many histories that you do not know that you are looking for. There is so much behind everything; for me, this is something to transmit in landscape design. To make the thoughts moving in space and time.

WU: There were two special aspects in this project. First, your office is nearby so that you understood the mountain very well and were involved from the very beginning. Second, I know you are a pilot. It seems to me that these two privileges had equipped you with a unique view about Cardada: you could see it from the sky and observe closely all its details. How did these two utterly different perspectives fulfill your design?

BÜRGI: Nobody has asked me this question, interesting; it seems to suggest that I may have had some "privileges of perception". Cardada project lasted five year and I surveyed many places in the mountain in person, so the convenience was certainly there. But design is mainly a way to feel and to approach. I work on many foreign projects often. It can help to be close to the place, but it can also help because you are far from the site, then you can see it with the eyes of a person who comes from abroad. Such views may be more open. They are the eyes of a curious observer. They are not the eyes of a person who is used to being there and thus takes thing for granted. In a recent project on the aesthetics of the urban agricultural land in Germany, we wandered on the site to see wild flowers. It was also very much related to the site and its history. We wanted to take up a respectful view and opinion of the farmers cultivating the land.

WU: We will return to this urban agricultural project a little bit later. You just mentioned a very interesting concept of the designer being a curious observer of the site. What is your entry point when encountering a new site?

BÜRGI: There are many points that bring you to the way you look at the site. Maybe the first approach is the form of the site, its structure, its topography, and so on. But then there are many other things to consider. For instance, there is light. Light can play a very important role. And then of course, there is history. If you are at a place and you know the history of this place, then this is the point that transmits other feelings. If it is a field — a mere piece of agricultural land — and maybe you find out that many centuries ago there was a battle on this site, then this field becomes something else. So it is a question of feeling, experience and mood. Anything inspiring you at the moment can give you some input to the design.

WU: Design criticisms often associate your work with minimalism. What is your opinion? Do you think your work is minimalist?

BÜRGI: Yes, there is the minimalist aspect; but in that transmits more. Design is a process. The kind of study that I do asks to take away and to eliminate; the given expression is therefore very strong. To be more precise, it is a process of distill; concise can only come from richness. This method of distillation is not only in architectural and landscape design, but also in music and literature, like composing a poem. It is really an achievement to be able to let aside everything that is not important. It is very difficult, demands long time training. I like conciseness, in art, in literature, and of course in design, because their expressions are strong, condensed and powerful.

WU: Many of us probably feel burdened by too many things in contemporary life. Do you intend in your design to eliminate everything that is not essential in order to help people reconnect with nature?

BÜRGI: I really think that you can transmit in each project a way of looking at your surroundings, a way to be more attentive and sensitive, even more respectful to the environment. As a designer, you need this capacity to eliminate but it is also important to try to transmit something. To transmit something, you need a feeling. Forming a drawing or a project, this comes from feeling, from the desire to transmit something. Of course, beauty is also very important. We notice, for instance, in the Cardada project, we have no vandalism. If somebody really likes something, he does not destroy it. If you notice where people spray and write on walls, often it is not on something that transmits beauty.

WU: I would like to hear your opinion of "beauty". It seems to be a word many designers nowadays are afraid of using. For example, your new project near Cardada, the "Giardino dei Percorsi".

BÜRGI: The Giardino Percorsi is a very small private garden. Let me first mention a detail: all my gardens have their own names, just like the naming of yuanlin, which do not follow the name of the client. Giardino dei Percorsi means a garden that has many ways of walking through. The villa locates on a slope overlooking the lake. Inside the garden, there was only a place to sit near the house. My design explores the different possibilities of walking through this space. The way you cross the lawn is important because it establishes a relationship between the horizon, the lake, and the islands. I think it is a particular pleasure if you have a place that you can walk through and experience different perspectives in and have certain places where you can just stop and sit. Or if you come to the edge of the property, when you stand there, you see something — a particular window in the landscape — or you see a particular tree in the woods or whatever. We extend the garden to the cliffy slope and planted some exotic plants that grow on the rocks. An experience of freely strolling in nature is what this garden is about. This is what I think about beauty. It is not a particular form or pattern, but the experience of feeling and discovering, of becoming a curious observer. That is fantastic and beautiful. That is garden to me.

WU: What do you want to transmit in this design?

BÜRGI: Many different atmospheres. For instance, the surprise of looking at the horizon, the pleasure of moving through the garden, or the way one arrives the space. Wherever you are, there must be an invitation to stop for a while to look and observe things.

I told you about my recent experience in China. I had visited Suzhou more than 20 years ago and still remember it clearly. Last year I wanted to see those gardens again. Someone warned me: "The previous impression may make a second visit tasteless." I admired some new gardens, but also those I've already visited before. Revisiting, I have not changed my opinion. On the contrary, after careful study I could enjoy them much more, appreciating how rich they are. For instance, the "Master of Nets" garden was one of my favorite gardens. It is such a garden that, how shall I say, you can never consume it. You can never totally experience it. Walking in it, there is always a landscape that you have never been opens in front of your eyes. You can keep on walking and walking. It is a process of discovery that never ends.

I appreciate Chinese gardens because these are works of art that can only be made after having been a deep observer for a long time. Another garden (The Mountain View with Embracing Beauty) is very small. At first sight, you see a central space with a composition of rocks divided by some water ponds. When entering, you will feel as if walking on a real mountain — not a small mountain, but a high mountain. A fantastic way to transmit all the experiences of climbing. In order to make a work like this you have to be a fabulous observer. Moreover, you must be able to give form to this feeling. Many have the experience of mountains. But how to give form to feeling and to make it into a garden, that is about design, about art. Do you know what is fantastic? These gardens are already centuries old, yet they still have the same energy. Many designs — those fragile once — only give superficial splendor, but their experience is ephemeral and short-lived. Art should be long-lasting.

WU: To read Chinese gardens as "giving form to the experience", this is indeed something many people missed when visiting the Suzhou gardens. They only see the form but not the experience behind it.

BÜRGI: Last year in China, I gave a lecture in Nanjing. I explained my philosophy — how I was very curious and my thoughts on Chinese art — so many audiences came to tell me that they shared so many things in common. I felt very proud to be able to share similar feelings with others. I was also fascinated when visiting the gardens with Chinese professors. We looked at these gardens like Europeans looking at art. It is not just about going through but about reading the gardens. There are so many things helping your thoughts: you walk, you see, you smell, you read, you hear... it is a total experience.

WU: One has to be a broad person in order to be a good designer.

BÜRGI: Right. I am curious about many things, things that happen everyday. I may get an experience from a sentence in a newspaper. I am glad when having answers to your questions... A story about Canadian pianist, Glenn Gould, inspired me: He was traveling on the highway— maybe in the States —and stopped in one of those roadside restaurants. Something all over the world, characterless. Around him are several tables of different travelers, talking and eating — you know, a very common scene. Then Gould starts to pay attention to what people are discussing, one table after another, totally different. Back to his studio, he composed a piece of music featuring different melodies that come and go. How boring and impersonal are those highway restaurants, but he uses it as a source of imagination, of creativity, and translates it into a piece of marvelous music. Why cannot we transmit daily life experiences into architectural and landscape designs? If only one is a curious observer, one can find inspiration and experience everywhere.

WU: One of your projects, Parco di Casvegno, is just like music.

BÜRGI: Yes, the project started 25 years ago. The story behind it is interesting. The site was a existing park land casually planted, with no design but a big surface. The neuropsychiatric hospital who owns the park wanted to open it to the public, but people normally would not like to walk around inside a hospital. So I imagined a park that would be totally away from the hospital building. Around this park I wanted to create some intervention — "follies" with trees. It was a very cheap project, cost nothing. The idea of "folly" may seem ironic. "Folly" means crazy and is the name of a type of decorative structure in traditional European gardens. I said to myself, "Am I allowed to do crazy things with trees in a place for mental sickness?" Then I had lunch with a psychiatrist, but I found out that he was crazier than I, so I decided to do what I want. Every year we did plan to add one object in this park. I must say I am surprised how powerful nature is.

WU: Is this project still going on?

BÜRGI: Yes, there are a total of seven or eight different tree follies in the park. They are still growing. Unfortunately, one of them has been destroyed in between.

WU: When visiting your office, I was surprised that you also own a huge nursery! Botanic knowledge is certain very important in creating these of garden follies.

BÜRGI: I also did projects without trees that were only concrete which have its own language. However, over the years, I find more and more meaning in the tree element of a project. A tree is not just something that you plant and grow. A tree has a history, a meaning. The meaning of a tree can change from one country to another. Some trees like to grow alone. Others can be planted very close to one another and eventually grow together. And others, there are fights between individual trees. In 1985, I planted a spiral with poplar trees for a land art project involving one hundred poplars. When first planted, the trees were all exactly the same size. Now they are all twenty-five years old and all different. Some strong, some skinny, a few are sick, a few died already. It is like a community. It is life itself.

WU: Can you talk about the urban agricultural program in Essen, Germany, which you mentioned earlier?

BÜRGI: The project began three years ago. Landscape design took two years and is finished now. It is a very different type of intervention, in order to transmit a view of the agricultural landscape. The question was, "can a farmer cultivate and think aesthetically at the same time?'" It is a very challenging and new question. There are few examples in history that can serve as a point of reference. A project like this has virtually never been done. For us, it was a research project that had a lot of different interventions that shared the common desire to transmit beauty but not at the cost of the production. We try to marry beauty and utility. We began with a few lines of colors in the fields.

This project is very subtle in the sense that it seems to be on the edge. You walk through this landscape and you see an intervention but you ask yourself, "Is it done by the man? Does it have to be? Why did they do it? Is it casual? Why is there a line of flowers?" It does not give you answers; it leaves you with questions — like the Cardada project. These interventions create a kind of curiosity, making urban population to look at agricultural landscape with a new sensibility. Also, another important point was that we must remember that agricultural landscape is the result of hard work. In order to make this visible, we researched how farmers worked with the machines, how they moved the soil, and how these can be expressed in an aesthetical way.

WU: How big is the land? What kinds of crops were planted? And were they mostly chosen for aesthetic reasons?

BÜRGI: The surface was about a hundred hectares. It is huge, on the scale of Central Park. We introduced different kinds of flowers into the crops. The design is subtle and detailed, demanding alarge amount of research. We learn many things from the farmers, study and understand how they work. For example, when it is time for harvesting and sowing, how we cultivate, and so on. Only after these problems are cleared, we started to intervene. Rows of flower will bloom between the wheat fields in various seasons.

WU: Are there somebody still cultivate the land?

BÜRGI: Yes, of course. The cultivation will be carried out as usual. This project do not follow the old pattern of transforming productive fields into recreational parkland, but to make visible the real beauty of an agricultural landscape through hard work. It is a totally new concept of urban park. The farmers will receive some subsidy from the park service to take care of the planting. Meanwhile, their cultivation lowers down the municipal maintenance cost. It is a very challenging project, in the sense how much it has demanded all participants to engage with the environment with new sensitivity, to be a curious observer. And the design is there to transmit the feeling and experience of nature, season, growth...

WU: This project demonstrates a perfect marriage between urban park and agricultural landscape, far more insightful than ordinary patterns followed in many cities. Had time permit, I would love to learn all the details. Apparently curiosity and observation have benefited your design greatly. However, such a wonderful way of drawing inspiration from almost everything in daily life cannot be achieved in one day. How do you train your students?

BÜRGI: I transmit to them my feelings, without any secrets. A professor may stick with your own discipline or encourage students to strike out and create in their own way. In most schools, most of the students are very good at researching, plotting, mapping, observing, and explaining the processes. But the step where you must give form to a feeling is always the most difficult for all students. For me, it is very challenging to work precisely on this step: that is, how and why do we go from perception to form. Of course, this is the most incredible step in the process: it transmits a feeling and gives form to a feeling — the most beautiful moment in the creative process.

WU: Would you suggest young students to train themselves in a certain way?

BÜRGI: When a student comes to this point, the professor can accompany him. I think it is a step such that once it enters into your mind, then you can work this way. I noticed for instance, when teaching in Italy, students really wanted background in history. But this kind of knowledge is sometimes so heavy to carry on your shoulders that you lose the capacity to create — the capacity to even be creative. I think it is important that students keep open minded and strike out from their discipline. You should be interested in anything: in music, art, architecture, and politics, in everything really. This curiosity can give you a kind of richness that you can translate into your projects. In fact, often the drawing that comes from the mind, through your arm, through the hand, and comes to the paper is a result of long research work and observing, but these processes cannot be constructed. Maybe it comes and maybe it does not. General knowledge is very important. I noticed that only a few people now are curious to know. I have noticed while teaching in Europe and also in the States that in many young people, their general knowledge is very reductive. I think we need these backgrounds for creative processes such as drawing. Even if you can not relate one body of knowledge to another, it is important that you know there is connections. The issue is the reinterpretation of feelings in designed form.

WU: You pointed out a very important issue: reinterpretation as a method for creativity. You have been teaching in Europe and America, and visited Chinese universities. Can you share your view about landscape architecture education?

BÜRGI: Certainly. My reinterpretation is not a reinterpretation of forms but rather a reinterpretation of a feeling or an experience. It is translation, if you wish. And I believe this is extremely for Chinese designers. China has such a rich culture. Everything is there. If I would work or teach in China, my focus would be on reinterpretation and translation. How to read and perceive the history fully and in a holistic way, and then find a contemporary language to express it is the key challenge. It is not merely copying of forms and description, but reinterpretation of value; it is to observe the culture and history of the land with greatest respect and receptive mind. Landscape architecture is an open field with infinite possibilities of creation.

Students need to be taught in the capability of feel, and then incorporate it in the design. Nothing is new in terms of our senses. The difficulty in design is how to express it through form. I love to work with students. They are able but need to be guided. It is important that the professor forces himself to enter the world of his students. I have seen students, in Venice, Philadelphia and Nanjing done beautiful researches on the site. The difficult is how to transmit the finding into the project. Design is an art, not merely analysis. I encourage my students to strike out, to extend their thinking, to keep open with their minds. Not to dive into refinement too soon, reexamine your ideas, be critique to yourself, and be ready to put everything upside down, all the time. And most important of all, be a curious observer of the world around you, of nature, of culture, of history, of society... these are where the real creativity issues from. (Transcribed by Xi ZHANG, Translated by Ying LI, Proofread by Tina TIAN, Xin WU)

附图：
被采访人及其设计团队作品展示

Appendix:
Interviewees and their design team project exhibit

001
俞孔坚 / Kongjian YU

1

2

附图：被采访人及其设计团队作品展示 / 203
Appendix: Interviewees and their design team project exhibit

3

1 天津桥园 © 土人景观
　Bridge Park, Tianjin © Turenscape

2 中山岐江公园 © 土人景观
　Qijiang Park, Zhong shan, Guangdong
　Province © Turenscape

3 秦皇岛滨海景观 © 土人景观
　Seaside landscape, Qinhuang Dao © Turenscape

6 明尼阿波利斯城市滨水概念设计 © 土人景观
Concept Design of Urban Waterfront, Minneapolis, Minnesota © Turenscape

4 上海世博后滩公园 © 土人景观
Houtan Park, Shanghai World Exposition 2010 © Turenscape

5 河北迁安三里河生态廊道 © 土人景观
Sanli River Ecological Corridor, Qian'an, Hebei Province © Turenscape

7 张家界哈利路亚音乐厅 © 土人景观
Hallelujan Concert Hall, Zhangjiajie, Hunan Province © Turenscape

002
戴安娜·巴摩里 / Diana BALMORI

1

1 艺术实践—曼哈顿史密森浮岛 © Balmori Associates
Artistic practice – Smithson Floating Island, Manhattan © Balmori Associates

2 线性景观—美国肯特瀑布,康涅狄格州 © Balmori Associates
Linear landscape – Kent Falls Trail, Connecticut © Balmori Associates

附图：被采访人及其设计团队作品展示 / 207
Appendix: Interviewees and their design team project exhibit

2

3

4

3 城市景观—西班牙比尔堡水滨 © Balmori Associates
Urban landscape—Campa de Los Ingless, Bilbao, Spain © Balmori Associates

4 MPPAT—韩国世宗特别市总体方案 © Balmori Associates
Master Plan for Sejong Public Administration Town, Korea – A Zero-waste Urban Plan © Balmori Associates

5 绿色屋顶—美国纽约百老汇屋顶花园 © Mark Dye
Green roof – 684Broadway Penthouse Roof Garden, Mankattan, NY © Mark Dye

附图：被采访人及其设计团队作品展示 / 209
Appendix: Interviewees and their design team project exhibit

5

003
贝尔纳·拉素斯 / Bernard LASSUS

1 COLAS集团屋顶花园：碧色剧坛中带有光之喷泉的洞室。碧色剧坛坐落在COLAS集团7楼的平台上，在公司董事局会议室对面。拉素斯被告知花园是禁止入内的，因此与其说它是个花园倒不如说它只是个花园的表象。拉素斯决定加强这种表象的印象。他创造了一个幽默的雕塑花园令人想到一个置于法国巴洛克花园中的露天剧场。台子上有一个花园的直白象征，即是一个花园洞室遮蔽着喷泉。碧色剧坛和洞室由着色并剪裁过的金属平板代表，给人一种视觉效果，很显然树与岩石的所在远不可及。我们在图片中看到洞室置于舞台之上，限于两个篱笆之中，由修短的树篱包围起来。洞室本身是一块巨大的红色和紫色岩石围着的一个黑黑的洞口，里面是彩色霓虹灯，令人想起悬在空中的喷泉。地面由大理石制成。在夜晚，花园中五彩变幻的喷泉看来想要挑战我们关于水的概念，而发出白光的神秘正方形投射出由光线组成的蓝色阴影，在黑漆漆的绿色屋顶上时隐时现，葡萄藤则自屋顶垂下，牢牢地固定在墙上。© Michel Conan

The Grotto with a Fountain of Light in the Green Theater Garden for Colas Group. The Green Theater Garden is located on a terrace on the seventh floor of the Colas Group in front of the corporation boardroom. Lassus was told that nobody would be allowed in this garden, so it would be a representation of a garden rather than a garden. Lassus decided to compound this image of representation. He created a humorous sculpture garden evocative of a Green Theater in a French baroque garden; with an on stage representation of a garden, symbolized by a garden grotto sheltering a fountain. The Green Theater and the Grotto are represented by flat metal slabs colored and cut out to create the visual impression of trees and rocks seen at such a distance that they are clearly out of reach. The Grotto that we see on this picture is placed on a stage limited by two yellow palisades and surrounded by cut hedges receding into the distance. The Grotto itself is a large rock of red and purple color surrounding a black opening within which colored neon lights evoke the presence of a water fountain suspended in the air. The ground is made of gravel. This image shows the garden at night with the changing color of the water of the fountain that seems to challenge our ideas about water, while a mysterious square of white light projecting a blue shadow made of light, appears from time to time in the dark hollow of a green cupola from which vines are dangling down in perfect immobility. © Michel Conan

附图:被采访人及其设计团队作品展示／ **211**
Appendix: Interviewees and their design team project exhibit

2

2 粉色小气象(Un Petit Air Rosé),1965 年。这是一件短暂的艺术作品,因为它经不住郁金香的寿命,即使拉素斯尽可能长时间地将白色纸片放在它的花冠里。然而它却成为拉素斯一生之中设计景观常用的灵感来源。事实上它是一个理论作品,在可能的范围内以抽象造型呈现出简单的图解,为后来的艺术发展提供了建议。拉素斯本人宣称,艺术作品有 3 个层次:作品本身、提供其形式的原型以及暗喻了原型的图示。© Atelier Bernard Lassus

A Little Rosy Air (un petit air rosé), installation by Bernard Lassus. This is an ephemeral art work, since it could not outlive the tulip, and even lasted only as long as Lassus held the piece of white paper inside its corolla. However it has remained a constant source of reference for Lassus throughout his life when creating landscapes. In fact it is a theoretical artwork, insofar as it presents in abstract fashion a schemata, which has given rise to later artistic proposals. Lassus himself asserts that there are three levels to an art work: the art work itself, the model that prefigures its form, and the schemata that anticipate the model. © Atelier Bernard Lassus

3 Guénanges 高中的彩色灌木丛，1972 年。这件艺术作品已经不存在了。一方面是构筑物，另一方面是草地上的树篱，它在两个极端中间扮演了一种视觉上的调和。© Atelier Bernard Lassus

The Colored Bushes at the School of Guénanges (1972). This art installation does not exist any longer. It created a visual mediation between the buildings at one extremity and the hedges of meadows at the other. © Atelier Bernard Lassus

附图：被采访人及其设计团队作品展示 / **213**
Appendix: Interviewees and their design team project exhibit

4

4 COLAS集团屋顶花园：四季园的小瀑布，人造草地和树，2002年。国际COLAS集团大楼最低的天台就坐落在集团会议大厅的外面，邻近有一间大房间用于艺术展览。这个平台将用于鸡尾酒会和隔壁房间活动后的接待处。就在外面低2m处，有一个植有树的公共花园。接待处木地板的外围是花园的元素。一条设计得就像高树篱的栅栏将其与公共花园分开，而栅栏的前面则放置着不同季节的树。在另外一侧，两个带轮子的花盆里面放满了抽象的花朵和草，让人想起草地。往左转你能看见由上百条水帘组成的小瀑布，空气中充满着溅落的水声。© Michel Conan

The Meadow, cascade and trees of the Garden of Seasons at the Colas Group (2002). The lowest terrace of the International Group Colas building is located outside of the Group conference hall, and near a large room devoted to art exhibitions. This terrace is meant for cocktail parties and receptions after an event in the neighboring room. Just outside, but a couple meters lower, there is a public garden planted with trees. The garden elements are all located on the periphery of a wooden floor used for receptions. It is separated from the public garden by a fence designed to look like a tall tree hedge, in front of which are placed the trees of different seasons. On the opposite side two garden planters on wheel are filled with abstract flowers and grasses evocative of a meadow. To the left you can see the cascade with its hundred waterfalls, which fill the air with the sounds of splashing water. © Michel Conan

5

5 装置艺术：松树照片系列。
  伦敦 Coracle 画廊，1981 年 1 月。
  © Atelier Bernard Lassus
  The Pine Trees Installation. Coracle Gallery London January 1981. © Atelier Bernard Lassus

6 COLAS 集团屋顶花园：俯瞰碧色剧坛的候望台上的人造树 © Michel Conan
  The Trees in the Waiting Room Garden, overlooking the Green Theater Terrace at the Colas Group © Michel Conan

004
帕特里夏·约翰逊 / Patricia JOHANSON

1

1 佩塔卢马湿地公园（加利福尼亚州）——投入运营使用后的水处理设施和公园景色 © Scott Hess / scotthessphoto.com, 2009
Petaluma Wetlands Park, CA—View of the site in 2008-2009 © Scott Hess / scotthessphoto.com, 2009

2 美丽公园潟湖（德克萨斯州）——场地总图。钢笔，33cm×66cm © Patricia Johanson, 1982
Fair Park Lagoon: Site Location, 1982. Ink, 33cm×66cm © Patricia Johanson, 1982

附图:被采访人及其设计团队作品展示/ **217**
Appendix: Interviewees and their design team project exhibit

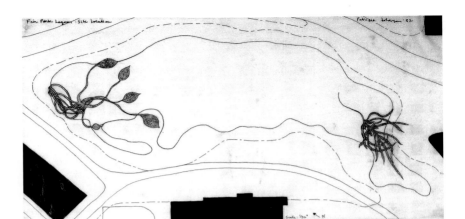

2

3 衣藻／景观设计／公园／雕塑,
钢　笔, 76cm×61cm © Patricia Johanson, 1974
Chlamydomonas / Landscape Design / Park / Sculpture, Ink on vellum, 76 cm×61cm © Patricia Johanson, 1974

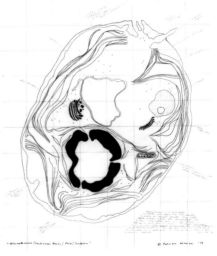

3

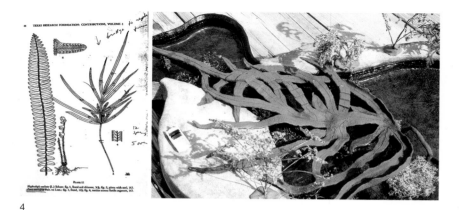

4

5

4 设计手稿—注页与植物图案，及设计模型—凤尾蕨雕塑（桥）© Patricia Johanson, 1982
Note-sheet: model for Fair Park Lagoon: Pteris multifida (Bridge) © Patricia Johanson, 1982

5 美丽公园潟湖（德克萨斯州）—扁叶慈姑景色，1982-1986。
喷浆，71.6m×71.6m×3.7m。
© Michel Conan, 2007.
Fair Park Lagoon—View of Sagittaria platyphylla, 1982-86. Gunite, 71.6m×71.6m×3.7m. © Michel Conan, 2007

附图：被采访人及其设计团队作品展示 / 219
Appendix: Interviewees and their design team project exhibit

6

6 美丽公园潟湖（德克萨斯州）—凤尾蕨景色，1982-1986。
喷浆，68.6m×34m×3.6m
© 吴欣，2007
Fair Park Lagoon—View of Pteris multifida, 1982-1986. Gunite, 68.6m×34m×3.6m. © Xin WU, 2007

7 佩塔卢马湿地公园（加利福尼亚州）
a. 以一种本地物种（盐沼鼠）为形的深度水处理池的设计图手稿 © Patricia Johanson, 2004；
b. 建设后的鼠形深度处理池鸟瞰图 © Herb Lingl / aerialArchives.com, 2005;
Petaluma Wetlands Park, CA-a. Sketch of the water polishing ponds in the form of a local species—the Salt Marsh Harvest Mouse © Patricia Johanson, 2004; b. Petaluma Wetlands Park,CA—Aerial view of the Mouse polishing ponds after construction © The city of Petaluma / aerialArchives.com,2005;

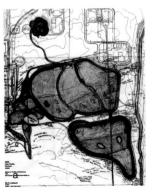 

7

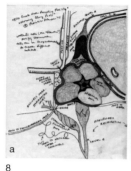

8

9

8 佩塔卢马湿地公园（加利福尼亚州）—— a. 牵牛花雨水径流净化池 © Patricia Johanson, 2004；b. 牵牛花雨水径流净化池选种了 16000 种有净化作用的湿地植物 © Scott Hess / scotthessphoto.com, 2007
Petaluma Wetlands Park, CA—a. Morning Glory Pools. © Patricia Johanson, 2004; b. Petaluma Wetlands Park—The Morning Glory Pools was planted with 16 000 wetland plants. © Scott Hess / scotthessphoto.com, 2007

9 佩塔卢马湿地公园（加利福尼亚州）—在深度处理池中引入特选的藻类来起生物氧化作用，将废水中的有机物氧化分解，达到净化目的。覆有水藻的巨型氧化池表面的颜色随水体净化的过程而变化，看起来就像是抽象画。最后，水中密集种植的一人多高的图莱草（Tule）把水藻过滤下来，与净水分离。
a. © Petaluma government, 2009; b.© Patricia Johanson, 2010；
Petaluma Wetlands Park, CA — Algae are introduced in the polishing ponds to oxygenate the water. The colors of these algae-covered ponds change with the seasons, looking like abstract paintings. Later, the algae will be filtered out by the bands of dense Tule planted in the water. a. © Petaluma government, 2009; b. © Patricia Johanson, 2010)

10 佩塔卢马湿地公园（加利福尼亚州）—鼠形深度处理池内的鼠眼也是功能性的导流岛。它能加快流速增进氧化作用，同时又是动物的筑巢地。远处山坡上的生态葡萄园是用这个水厂生产的循环水浇灌的。© Scott Hess / scotthessphoto.com, 2008
Petaluma Wetlands Park, CA—View of an island inside the polishing ponds The island, which directs the flow of water through the sewage treament pond, is also the nesting place for animals. The vineyards across the highway are irrigated with the recycled water from the facility, © Scott Hess / scotthessphoto. com, 2008

005
埃里克·董特 / Erik DHONT

1

1 埃里克·董特的成名作:与一座建于1602年的城堡和谐共存的现代抽象曲径迷宫。比利时,盖斯比克 © Jean-Pierre Gabriel
Dhont's first remarkable design: a contemporary abstract labyrinth stands in harmony with a 1602 castle, Gaasbeek, Belgium © Jean-Pierre Gabriel

2 新近建成的位于葡萄牙阿索里半岛圣·米盖尔的植物园。© Jean-Pierre Gabriel
A newly built private botanic garden in The Azores, Portuguese. © Jean-Pierre Gabriel

附图：被采访人及其设计团队作品展示 / **223**
Appendix: Interviewees and their design team project exhibit

2　　　　　　　　　　　　3

4

3 以欧洲造园传统中的树木修剪法为蓝本，为一位当代艺术收藏家的花园设计的银杏围廊和树雕小径。比利时，奥亨。© Erik Dhont
"Ginko-pallisade" and "topiary-walk" for the garden of a contemporary art collector, based on the European garden tradition of topiary, Ohain, Belgium. © Erik Dhont

4 埃里克·董特的抽象设计图
© Erik Dhont
Dhont's abstract design drawings © Erik Dhont

5 景观空间研究模型和树雕造型的雕塑性形体实例。© Reiner Lautwein
Studies of spatial relationship and topiary form. © Reiner Lautwein

6 新近建成的位于美国旧金山玛拉埠海湾的私家雕塑园。© Jean-Pierre Gabriel
Newly completed sculpture garden, Malibu (San Francisco), USA. © Jean-Pierre Gabriel

附图：被采访人及其设计团队作品展示／
Appendix: Interviewees and their design team project exhibit

7

8

7 为北京尤伦斯当代艺术中心所做的设想性"一英里"长条状公园研究，塑性体量与乔灌木有机形体相结合的景观空间。© Reiner Lautwein Proposal for Ullens Center of Contemporary Art garden: one-mile long garden with sculptural forms and organic tree shapes, Bejing, China. © Reiner Lautwein

8 将回收的旧砖成功地引进到为城市新潮族的住宅景观设计中，比利时，劳申莱尔。© Jean-Pierre Gabriel
Introducing recycled bricks into chic urban residence, Roesselare, Belgium. © Jean-Pierre Gabriel

附图：被采访人及其设计团队作品展示 / **227**
Appendix: Interviewees and their design team project exhibit

9

9 埃里克·董特工作室利用建筑废料建造的3个低造价"社会性花园"
a.b. 位于比利时布鲁塞尔大科第奥斯博物馆瀑布模型和建成景观 © Erik Dhont, Jean-Pierre Gabriel
c. 位于比利时布鲁塞尔自然野趣的梯步道
© Jean-Pierre Gabriel
d. 位于比利时鲁汶回收材料的石地毯
© Jean-Pierre Gabriel

Three "social gardens" built of recycled materials by Erik Dhont Office:
a.b. Fountain at Musée Le Grand Curtius, Brussels, Belgium: model and built view © Erik Dhont, Jean-Pierre Gabriel
c. Stair walkway with input of nature, Brussels, Belgium © Jean-Pierre Gabriel
d. Stonecarpet with reused materials, Leuven, Belgium © Jean-Pierre Gabriel

006
林璎 / Maya LIN

1

1 北纬10度
（诺曼·麦克格雷斯摄影；佩斯画廊）
10° North
(Photography by Norman McGrath.
Courtesy The Pace Gallery.)

附图：被采访人及其设计团队作品展示 / **229**
Appendix: Interviewees and their design team project exhibit

2

3

2 斯托金波场
  （杰瑞·L·汤姆森摄影；佩斯画廊）
  Storm King Wavefield
  (Photography by Jerry L Thompson.
  Courtesy The Pace Gallery.)

3 读花园
  （林璎工作室摄影；佩斯画廊）
  Reading a garden
  (Photography by Maya Lin Studio.
  Courtesy The Pace Gallery.)

4

4 输入
(布莱恩·布劳斯摄影;佩斯画廊)
Input
(Photography by Brian Blauser.
Courtesy The Pace Gallery.)

5 食
a. 日景(巴尔萨扎·克拉柏摄影;费莱基金会)
b. 夜景(苏其特拉·凡摄影;佩斯画廊)
Ecliptic
a. Day time view(Photography by Balthazar Korab. Courtesy Frey Foundation.)
b. Night view(Photography by Suchitra Van. Courtesy The Pace Gallery.)

附图：被采访人及其设计团队作品展示／ **231**
Appendix: Interviewees and their design team project exhibit

5

007
帕奥罗·伯吉 / Paolo BÜRGI

1

2

1 卡尔达达平面示意图：地质观测坪平面显示岩石标本和欧非板块的分界线（图中红线）
© Paolo Bürgi
Cardada (Locarno, Switzerland): Plan of the geological observatory with the Insubric line and the placement of the rock samples of the European and African plate © Paolo Bürgi

2 竹径（瑞士私家花园）：竹径尽头的休息平台，悬在野趣盎然的不可知的自然中
© Paolo Bürgi
Bamboo Path(private garden, Switzerland) The small terrace revealed at the end of the Bamboo Path is suspended over a wild, unapproachable landscape © Paolo Bürgi

附图：被采访人及其设计团队作品展示／ **233**
Appendix: Interviewees and their design team project exhibit

3 佩科西花园／万径游园 （瑞士私家花园）
© Paolo Bürgi
Giardino dei Percorsi / Garden of the Paths (private garden, Switzerland) © Paolo Bürgi

3

4

5

4 城市农业（德国埃森市）© Paolo Bürgi
Urban Agriculture (Essen, Germany) © Paolo Bürgi

5 声景观（瑞士）：一个关于城市环境中声音的研究项目。© Paolo Bürgi
Soundlandscape (Switzerland): a research work about sounds in urban contexts. © Paolo Bürgi

6 伽斯瓦涅公园（瑞士）：绿色树亭
© Giosanna Crivelli
Parco di Casvegno (Switzerland): Green follies © Giosanna Crivelli

7 卡尔达达的地质观景坪 © David L. Hays, 2006
Cardada, Geological Observatory Panorama © David L. Hays, 2006

8 卡尔达达的景观瞭望台
© David L. Hays, 2006
Cardada, Landscape Promontory © David L. Hays, 2006

9 城市农业（德国埃森）／景致与功用。一个注重于城市中农业用地的美学的研究项目。两年中发展起来的季节性景观与农事日历相呼应。© Paolo Bürgi
Urban Agriculture (Essen, Germany) / vista and utility. A Research project on aesthetics of agricultural landscape in urban context. Impression of the seasonal phases developed over a period of two years and interacted with the agrarian calendar. © Paolo Bürgi

"十一五国家重点图书"——《风景园林手册系列》

1. 风景名胜区工作手册（上下册）	（20180）	128元
2. 公园工作手册	（17106）	58元
3. 风景园林师设计手册	（20375）	68元
4. 园林施工手册与养护手册	（20017）	42元
5. 城市园林绿化管理工作手册	（17480）	45元
6. 园林设计树种手册	（15119）	148元

建筑文化图书中心是隶属于出版社的专业图书中心，出版范围覆盖古代与现当代城市文化、建筑文化、建筑室内设计与装修、装饰工程与材料、家具与陈设、建筑旅游、风景名胜旅游、建筑摄影、风光摄影等方向，致力于大建筑文化的宣传推广工作，旨在为读者奉献具有建筑人文品格的精品出版物。我们拥有专业、敬业的编辑，有经验丰富的出版和营销人员，有爱书、爱建筑和爱文化的团队，我们希望和您一起共同为建筑文化的传播贡献一份力量。欢迎各位专家踊跃赐稿，联系电话：010-58337173；邮箱：wangxu@cabp.com.cn

图书在版编目（CIP）数据

景观启示录——吴欣与当代设计师访谈／吴欣编著．
北京：中国建筑工业出版社，2012.7
　ISBN 978-7-112-14247-7

　Ⅰ．①景…　Ⅱ．①吴…　Ⅲ．①景观设计-文集　Ⅳ．
①TU986.2-53

中国版本图书馆CIP数据核字（2011）第077243号

策划：孙立波
责任编辑：郑淮兵
责任设计：李志立
责任校对：肖　剑　王雪竹

景观启示录
——吴欣与当代设计师访谈

The New Art of Landscape
— Conversations between Xin WU and Contemporary Designers

吴　欣　编著

*

中国建筑工业出版社出版、发行（北京西郊百万庄）
各地新华书店、建筑书店经销
北京方舟正佳图文设计有限公司制版
北京云浩印刷有限责任公司印刷

*

开本：880×1230毫米　1/32　印张：7　插页：18　字数：251千字
2012年8月第一版　2012年8月第一次印刷
定价：30.00元
ISBN 978-7-112-14247-7
　　　（22320）

版权所有　翻印必究
如有印装质量问题，可寄本社退换
（邮政编码 100037）